African Violets and Other Flowering Houseplants

Created and designed by the
editorial staff of Ortho Books

Edited by Barbara Ferguson

Coordinating Editor Pam Peirce

Written by Dr. Charles Powell, Jr.

Designed by Ron Hildebrand

Paintings and drawings by Ronda Hildebrand

Ortho Books

Publisher
Robert L. Iacopi

Editorial Director
Min S. Yee

Managing Editors
Anne Coolman
Michael D. Smith
Sally W. Smith

Production Manager
Ernie S. Tasaki

Editors
Jim Beley
Susan Lammers
Deni Stein

System Manager
Christopher Banks

System Consultant
Mark Zielinski

Asst. System Managers
Linda Bouchard
William F. Yusavage

Photographic Director
Alan Copeland

Photographers
Laurie A. Black
Richard A. Christman

Asst. Production Manager
Darcie S. Furlan

Associate Editors
Richard H. Bond
Alice E. Mace

Production Editors
Don Mosley
Kate O'Keeffe

Chief Copy Editor
Rebecca Pepper

Photo Editors
Anne Pederson
Pam Peirce

National Sales Manager
Garry P. Wellman

Sales Associate
Susan B. Boyle

Operations Director
William T. Pletcher

Operations Assistant
Gail L. Davis

Administrative Assistant
Georgiann Wright

Address all inquiries to
Ortho Books
Chevron Chemical Company
Consumer Products Division
575 Market Street
San Francisco, CA 94105

First Printing in April, 1985

1 2 3 4 5 6 7 8 9
84 85 86 87 88 89

ISBN 0-89721-042-5

Library of Congress Catalog Card
Number 85-060002

Consultants:

Dr. Elizabeth McClintock
Sylvia Peterson,
J.A. Peterson Sons Greenhouse

Photo Credits:

Laurie Black: Front cover, Title page, pp. 6, 7R, 10L, 10R, 11L, 11R, 16, 17, 18, 19, 37, 38, 40, 45TL, 45TR, 45B, 46T, 46B, 47T, 47B, 50, 52, 53, Back cover TL and BL.

Richard Christman: pp. 23C, 26L, 26C, 26R, 27TR, 27B, 28, 43L.

Josephine Coatsworth: pp. 9, 44, 76, 79.

Alan Copeland: pp. 20, 43R, 51, Back cover BR.

Tony Howarth: p. 7BL.

Michael McKinley: pp. 65BR, 66L, 66TR, 66BR, 67L, 67R, 68R, 69R, 70L, 70TR, 71TR, 72L, 72TR, 72BL, 73TL, 73TR, 73CR, 73BL, 73BR, 74L, 74R, 75C, 75R, 77L, 78R, 79C, 79R, 80L, 80C, 81R, 82L, 83L, 84, 85L, 85R, 86TL, 86BL, 86R, 87R, 88L, 88R, 89L, 89C, 89R, 90L, 90R, 91L, 91R, 92L, 92TR, 92BR, 93L.

Phil Meyer/Black Star: pp. 61B, 63BL, 65TL.

Jack Napton: 81L.

David Papas: Back cover TR.

Pam Peirce: pp. 7TL, 15, 24T, 24B, 25T, 25BL, 25BR, 27TL, 29, 32T, 41T, 41B, 56TL, 56TR, 56BL, 56BR, 57T, 57BL, 57BR, 58TL, 58TR, 58B, 59T, 59B, 60, 60BL, 60BR, 61, 62, 63TL, 63TC, 63TR, 63BL, 64TL, 64TR, 64B, 65TL, 65CL, 65BL, 68L, 69L, 70B, 71L, 72BR, 75L, 77R, 78L, 80R, 82R, 83R, 87L, 92C, 93R.

Barry Shapiro: p. 39.

M. Shurtleff: p. 23T.

Ortho Photo Files: 21, 23B, 30TL, 30TR, 30CL, 30CR, 30BL, 30BR, 31T, 31C, 31B.

Designers:

Carolina West Designs, San Francisco, CA: pp. 11R and Back cover TL.

CoClico and Co. Inc., Interior and Exterior Landscape Design, San Francisco, CA: p. 38.

Cois Pacoe Design, Sausalito, CA: Front cover, pp. 10L, 18, 19, 47B, Back cover BL.

Helen Craddick, Cole Wheatman Interior Designers, San Francisco, CA: 47T.

Nancy Glenn Designs, Sausalito, CA: pp. 16, 17, 37, 46B, 53.

Glenn Design Co., Sausalito, CA: 52.

Geoffrey Gratz, San Francisco, CA: 7R, 40.

The Kreiss Collection, San Francisco, CA: pp. 11L, 45TR.

Ruth Livingston, Tiburon, CA: pp. 6, 46T.

Angel Tardy/Interior Plant Systems, San Francisco, CA, 45TL.

Chevron Chemical Company
575 Market Street, San Francisco, CA 94105

Special Thanks to:

African Violet Growers: Pearl Borden, Donna Burla, Barbara Elkin, June Fallaw, Dorothy Gawienoski, Fred Grafelman, Ted Khoe, Larry Lau, Harriet Poss and the African Violet Society Chapters of San Francisco and San Jose. Nurseryman's Exchange, Half Moon Bay, CA; J.A. Peterson and Sons Nursery, Cincinnati, OH; Sunnyside Nurseries, Fremont CA.

Others: Pamela Bain; Tim Brydon; Conservatory of Flowers, Golden Gate Park, S.F., CA; William D. Ewing; Ferry Morse Seed Co., San Jose, CA; Kathi Gonzalez; Kew Gardens Conservatory, Kew, Surrey, England; Marc Merman of the Ornamental Horticulture Dept., City College of S.F.; Dr. Earl Murphy, Oakland, CA; Natural Resources Library, University of California, Berkeley; Ornamental Horitculture Department, City College of S.F.; Sharon Supon, Strybing Arboretum, Golden Gate Park, S.F., CA; Wintergreen Nursery, Watsonville, CA.

Self-Watering Devices:

Pot: J. & J. Swift, Box 28012, Dallas, TX, 75228. *Saucer:* Optiflora Inc., Box 8158, Nashville, TN, 37207. *Siphon:* Blumat, Edmond Scientific Co., 101 E. Gloucester Pike, Barrington, NJ, 08007.

Front cover: African violets and rabbit's foot fern make a living flower arrangement.

Back cover:

Top left: White azaleas add snowy charm to a warm indoor setting.

Top right: African violets may be grouped to decorate a formal dinner table.

Bottom left: Persian violets are delicate, sweet scented, and long lasting.

Bottom right: Set alone on a black marble table, African violets steal the show.

African Violets and Other Flowering Houseplants

The art of growing flowers in the home has been pursued for centuries, and today homes are better adapted than ever to meet the needs of these special houseplants. Whether plants flower indoors continuously, bloom seasonally, or are brought indoors only during the blooming period, they beautify the home.

The Flowering Houseplant

Growing and displaying African violets and other flowering plants in the home is a natural evolution of the enormous interest in the houseplant hobby. The diversity of plant types available to choose from, the opportunity to challenge and increase your plant growing skills, and the exhilaration that comes from the colors, scents, and variety of blossoms are all inescapable attractions for those with an interest in indoor gardening.

Cut flowers have always been popular decorations for homes. Over the past few years, as more and more people learn to think of flowers as a necessity rather than a luxury, the demand for cut flowers has increased tremendously. The cultivation of flowering indoor houseplants combines the excitement, beauty, and decorative effect of cut flowers with the permanence and vigor of healthy houseplants.

If you want to grow a flowering houseplant, you will need skill and an understanding of the plant and its needs, but you will be rewarded when you watch the delicate flowers unfold, adding charm and a touch of spring to your room. Once blooming, a living plant often continues its display for weeks or months, thus offering much greater value than a short-lived cut bouquet.

A primary purpose of this book is to help you grow your African violets to perfection. African violets are the best known of the flowering indoor plants. In recent years, there have been many changes in the way they are bred and developed. These changes, along with their associated changes in cultural practices, have made these popular plants even more widely available and easy to grow. Perhaps you've tried them before and have felt they could perform better. Perhaps you've had no luck in the past and are determined to try again. Maybe you've seen them as bright centerpieces in friends' homes and have finally decided to grow your own. Whatever the reason, the instructions and guidelines for African violet culture described in this book will give you a better understanding of these plants and will enable you to fit them into your home and life-style.

But why stop with African violets? In many cases, the basic guidelines for growing these plants are identical to the culture of other plants that grow and flower indoors. In this book you will also be introduced to the ever-expanding selection of indoor flowering plants and their culture and use.

Finally, this book will help you learn more about caring for and displaying flowering plants indoors. Most flowering houseplants are brought into an indoor setting already in bloom. Seasonal plants, such as poinsettias, chrysanthemums, and various bulbous plants, are enjoyed by millions of homeowners and apartment dwellers throughout the country. Although these plants are grown and purchased for temporary use indoors, they can be cared for in a way that will immeasurably increase their beauty and longevity in your home. As for the flowering plants that you keep in your home the year around, this book will give you the information you need to bring them into bloom and keep them blooming for an extended period of time.

In the Chapters of This Book

Although the emphasis is on African violets, the chapters in this book are not overly specific to this group. African violets are a good flowering houseplant to begin with, especially now that the cultivars are so vastly improved; but you can apply your newly acquired expertise to the wide assortment of flowering plants available as indoor houseplants.

In this first chapter, we will go on to explain a bit about the kinds of plants included in the definition of "flowering" houseplants. Once you have read about the different types of plants that flower indoors, you can decide for yourself which ones you are most interested in selecting, growing, and displaying in your home.

In the second chapter, "Growing Flowering Plants Indoors," you will learn how to bring plants into flower indoors and prolong their blooming period. To do this, you need a basic understanding of plants and how they function. Light plays a key role in the growth and health of the plants, and modern indoor lighting techniques have made it possible to grow and display a wide range of flowering plants indoors.

This chapter explains plant processes and how they depend on light to function. Indoor plants also need water and nutrients, temperature and humidity, and pots and potting mixes for proper growth. When you know how to provide your plants with the correct environment for maximum growth and blooming, this chapter will show you ways to propagate new plants from your present collection.

You may find, after reading these sections, that your general horticultural or gardening knowledge is quite good. The tricks to becoming a successful indoor flowering plant gardener are merely extensions of ideas that come from a well-founded understanding of how to provide for the needs of plants. The novice gardener should combine this information with the basic principles contained in Ortho's book *All About Houseplants*.

The third chapter, "The Window Garden," is about window gardens and light gardens. Whether you have a bay window, a greenhouse, or an electrically lighted garden, you must design the space to maximize the available light and to enhance the beauty of the area. Here you will find ideas for arranging your plants in these growing stations, and instructions for creating an electrically lighted garden that is both useful and efficient.

Interior decoration with flowering houseplants is not only an art, it is also a highly regarded professional achievement. Designing with flowering plants indoors is the subject of the fourth chapter, "Displaying Flowering Plants Indoors." Its purpose is to help you to present flowering plants effectively indoors, to integrate them into the style of your home, and to use them to enhance your living environment.

In the last chapter, the encyclopedia of indoor flowering plants is a storehouse of information. It begins with the history and development of African violets, then gives a description of all the different types of African violets and their cultural needs.

The discussion of African violets is followed by a list of more than 80 other flowering plants that are widely available and are suitable for growing or for use indoors. Complete cultural information and environmental requirements, descriptions of growth and flowering habits, and photographs of many of the varieties will help you select the plants best suited to your home and life-style. The care requirements are specifically designed to enable you to bring plants into flower and to keep them flowering for as long as possible. The plants are divided into four categories: plants that bloom continuously, plants that bloom seasonally or intermittently, plants that need a rest period after blooming, and plants used indoors only while in flower. These categories are explained further on pages 9 and 10, and a list of selected plants in each category is given on page 11.

Flowering Plants in the Home

Above all, this book is about plants with flowers. Whether you grow them indoors, to eventually and periodically produce flowers, or you bring them

The pink flower bracts of the bromeliad Aechmea fasciata *make a handsome and long-lasting display.*

indoors when they are already in flower, this book will help you display and enjoy living plants with flowers in your home.

As you grow and bring into flower a wider and wider array of plants, the need for varied indoor environments becomes clear. Only a few years ago, most flowering houseplants were seasonal in availability and usefulness. Almost all had to be grown in a greenhouse, window box, or solarium to bring them into proper bloom and form. Such is still the case with many—but certainly not all—of the flowering houseplants available.

Today, many homes are constructed to bring natural light indoors. Concern about conserving energy has resulted in the construction of homes with large skylights, attached greenhouses, or passive solar collectors. New types of electric lights and improved installation techniques now give us the opportunity to grow more living plants indoors. It is indeed a green revolution in our life-styles and environments.

GROWING STATIONS AND PRESENTATION SITES

Many African violet enthusiasts grow their plants in a growing room or in the basement under fluorescent lights. These areas, set aside specifically for growing the plants and bringing them into bloom, are called "growing stations." Once the plants are fully grown and begin to flower, they can be removed from the growing station and enjoyed throughout the home. They can decorate the dining room table, cocktail table, or even a bookshelf, and they are especially handy when company is coming. But their stay in these "presentation sites" should not exceed one or two weeks (depending on the light and temperature conditions). In general, plants stop blooming when left in sites that were selected purely for decorative purposes. Most of the time the new buds die or develop inferior blossoms. It takes up to two months for African

Flowering plants can be displayed temporarily wherever you want them.

Plants can be rotated from a light garden (left), or special features may allow a permanent blooming plant display (right).

violets to bloom again or recover from an extended stay in a dimly lighted presentation site, even if they are returned to their optimum growing environment in a growing station.

Presentation Sites and Growing Stations as One

"The Window Garden" talks about light gardens, window boxes, and skylight settings. These indoor gardens are both presentation sites and growing stations. They are decorative, prominent displays that command attention, while also supplying all the cultural needs of the flowering plants.

Indoor Plants Through History

Interest in bringing the beauty of nature into the home is hardly a recent phenomenon. The practice has come into style and fallen out of style periodically over the centuries. Archaeological finds from Minoan times, before 1100 B.C., include many beautifully made plant containers — complete with holes in the bottoms! The Romans used their engineering skills to construct greenhouses with roofs made of huge sheets of transparent mica. The structures were heated by centralized furnaces with hot-water distribution systems.

At the beginning of the Renaissance, Venetian and Genoese merchants began commercial trade in plants, selling such Oriental ornamentals as hibiscus and jasmine to wealthy Europeans. It soon became fashionable to grow citrus and other tropical fruits in tubs. These were grown outdoors in the summer and overwintered in sheds, which soon gave way to glass-enclosed "orangeries," the forerunners of our modern-day greenhouses.

An exuberant Victorian tray garden combining plants and caged birds.

These greenhouses were unusual, and were constantly shown off to visitors and house guests. Thus more ornamentals — especially flowering plants — were brought into indoor culture. Soon the conservatory was born. A conservatory is a grand greenhouse built as much to display plants as to grow them. Early conservatories were often connected to the home. In this regard, they were not unlike the small greenhouses that are attached to many homes today.

The Victorian age of the nineteenth century brought about the tremendous popularization of indoor gardening. Families had more money to spend, and flowering plants were often used to create a luxurious indoor decor. The natural fragrances and colors of flowers were absolutely necessary in the boudoir of a fine lady. Although popular horticultural writings of the time stressed the need for adequate light in order to get plants to bloom, not everyone could afford a conservatory. Thus came window boxes, bay windows, and even the occasional skylight (usually made with translucent glass, rather than clear glass, no doubt for reasons of Victorian modesty.) Electrically lighted indoor gardens are a relatively new development. The first electric incandescent lights were generally unsuited for plant lighting because they generated far too much heat. If they didn't burn the plants, they burned themselves out — generally after only a few hours. In addition, the quality of light they emitted was not very efficient for plant growth.

Fluorescent lamps, developed in the late thirties, emitted reasonably high intensity light at cool temperatures. It was (and is) difficult to project or reflect such light, because it spreads out from its source, diminishing its effectiveness. However, this disadvantage has been

Although they can be grown indoors, huge draping vines and water lilies are subjects best left to large conservatories.

overcome with lamps of different hues and well-designed receptacles. In addition, the quality of light emitted by fluorescent lamps is always improving, making them more and more suited to plant growth.

Modern indoor gardeners have many advantages over those who lived in the Victorian age. Temperatures are more easily stabilized (despite our high heating costs) and our windows today are bigger, even in the most modest apartment. The soot-laden, smoky pollution of nineteenth-century cities has been largely eliminated. Even our water is of better quality, especially in major cities.

What Is a Flowering Houseplant?

Most people probably have their own definition of a "flowering houseplant." Aside from the fact that such plants have or are expected to have a flower, these definitions will undoubtedly be quite varied.

For the purpose of this book, let's agree on one definition. First, you are dealing with a houseplant. Whether they are selected for foliage or flowering habit, houseplants must be able to be grown and trained into a size, shape, and form compatible with the indoor spaces and living habits of the occupants of those spaces. Massive plants or those that flower only after many years of maturity are not included in this definition. Of course, such plants can be seen in greenhouses and were popular in the past in spacious, high-ceilinged plant conservatories. I can remember as a child visiting the local florist, who had a huge banana tree, lemon tree, and bougainvillea growing in a truly wonderful conservatory jungle. Such plants are beautiful, but they are rarely practical as flowering houseplants.

Aside from the overall size and shape of plants, you must include in your definition of flowering houseplant a differentiation between plants that grow and flower in a truly indoor living environment and those that require a greenhouse setting to initiate blooms. Most flowering houseplants adapt or acclimate to indoor light intensities. These light intensities will vary seasonally, but are usually one half or less of the intensity of light provided in a properly constructed greenhouse. Those plants that do require a greenhouse setting to initiate flowers can still be enjoyed indoors while they bloom, and we will include them within this definition.

USEFUL CATEGORIES FOR FLOWERING INDOOR PLANTS

Let's consider the "flowering" part of our houseplant definition in a bit more detail. Plants have a variety of methods for producing flowers. The method of flowering that a particular plant develops is closely tied to the native habitat of the plant and the plant's adaptation for survival. While it is not necessary to understand why a plant flowers the way it does, it is necessary to know how it flowers. For this reason, the plants described in this book are divided into four categories, determined by the ways in which they flower.

Plants That Bloom Continuously

Many African violets, begonias, and other plants grown indoors flower continuously, often for many years. These plants must be given optimum care in a stable setting. They are truly the best of the indoor flowering plants.

Plants That Bloom Seasonally or Intermittently

A wider variety of plants, such as peace lilies, Christmas cactus, or miniature geraniums, flower seasonally or intermittently throughout the year. They are induced to flower through proper management of temperature, light intensity, light quality, day length (or "night length"), and pruning practices. Their usefulness in indoor gardens is based on their attractiveness as foliage plants as well as flowering plants. These plants are often moved about the home, depending on whether or not they are in flower. Such practices are to be encouraged. However, indoor gardeners must realize that these plants may not continue to bloom if moved to a new, low-light location in order to better display the blossoms. In some cases, the plants may actually decline or become sickly if left for too long in a strictly decorative location.

Plants That Need a Rest Period After Blooming

Another group of indoor flowering plants are those that bloom seasonally or intermittently, but require a rest period (generally after they flower). Florist's gloxinias are the most widely known in this category. This plant goes completely dormant, the foliage dies back, and the tuber is removed and stored in moist sphagnum moss until the next year. With other plants (such as flowering maples and most bromeliads) the foliage remains, but the plants do not produce any new growth during their semidormancy.

Bulbous plants forced indoors are included in this group of plants that need a rest period. Although some bulbous plants are adapted for years of indoor culture, most of them wither and go dormant after their flowering period. This can present a problem for indoor gardeners. The foliage of most bulbous flowering plants is not particularly attractive. Yet, these plants must be given good growing conditions after they flower to provide a store of food for next year's flowering cycle. The answer is to move them to a presentation site during their flowering period. Prior to and after flowering, keep them in a growing station; a hidden spot in a well-lit window or greenhouse will do—but do not neglect them!

Plants Used Indoors Only While in Flower

Many bulbous plants are best placed in this final group of flowering houseplants: plants used indoors only while in flower. The most popular flowering houseplants are found here. Tulips, narcissus, poinsettias, chrysanthemums, and Easter lilies are often grown or forced in greenhouses by professionals and at the height of their flowering cycle. Such plants are presented rather than grown in the home. The satisfaction they give you is based upon their health when you buy them, your skill in presenting them well, and your ability to provide their minimal environmental needs while they are in flower. After the flowers fade, flamingo flowers, zebra plants, camellias, and gardenias grow well in a bright greenhouse, window box, or solarium; poinsettias, chrysanthemums, or cinerarias may best be discarded after flowering. Still others, such as crocus, narcissus, some tulips and azaleas, and Easter lilies, can be moved into the outdoor garden if the climate in your area permits this.

In general, plants used indoors only while in flower require no fertilization while they are being displayed in the presentation site. They do, however, need to be kept evenly moist and out of dry, hot drafts. Finally, an indoor setting that is too dimly lit will shorten the lives of these plants drastically. Although they will not die, they may drop some of their leaves, turn yellow, and the flowers may fade or fail to open fully.

Left: Rieger begonias bloom the year around in a suitable environment.
Above: Cool nights trigger seasonal bloom of cymbidium orchids.

Indoor Flowering Plant Categories

PLANTS THAT BLOOM CONTINUOUSLY INDOORS

Begonia × *hiemalis*—Rieger begonia
Begonia × *semperflorens-cultorum*—wax begonia
Columnea species—columnea
Episcia species—flame violet
Exacum affine—Arabian violet
Impatiens wallerana—impatiens
Nematanthus species—nematanthus
Saintpaulia species—African violet

PLANTS THAT BLOOM SEASONALLY OR INTERMITTENTLY

Aeschynanthus species—lipstick plant
Agapanthus species—lily-of-the-Nile
Ananas species—pineapple
Begonia × *cheimantha*—Christmas begonia
Bromeliads
- *Aechmea* species—living vase plant
- *Billbergia* species—vase plant
- *Guzmania* species—guzmania
- *Tillandsia* species—blue torch
- *Vriesea* species—vriesea

Daphne odora—winter daphne
Hoya species—wax plant
Orchids
- *Brassia* species—spider orchid
- *Cattleya* species—cattleya
- *Cymbidium* species—cymbidium
- *Dendrobium* species—dendrobium
- *Odontoglossum* species—lily-of-the-valley orchid
- *Oncidium* species—butterfly orchid
- *Phalaenopsis* species—moth orchid

Pelargonium species—geranium
Schlumbergera species—Christmas cactus
Spathiphyllum species—peace lily
Streptocarpus species—Cape primrose

PLANTS THAT NEED A REST PERIOD AFTER BLOOMING

Abutilon species—flowering maple
Achimenes species—rainbow flower
Begonia × *tuberhybrida*—tuberous begonia
Bulbous Plants
- × *Amarcrinum memoria-corsii*—crinodonna
- *Clivia miniata*—kaffir lily
- *Crinum* species—Bengal lily
- *Crocus* species—crocus
- *Eucharis grandiflora*—Amazon lily
- *Haemanthus* species—blood lily
- *Hippeastrum* species and hybrids—amaryllis
- *Hyacinthus orientalis*—hyacinth
- *Lachenalia* species—Cape cowslip
- *Lilium* species
- *Muscari* species—grape hyacinth
- *Narcissus* hybrids and cultivars—daffodils
- *Ornithogalum* species—star-of-Bethlehem
- *Scilla* species—squill
- *Sprekelia formosissima*—Aztec lily
- *Tulbaghia fragrans*—society garlic
- *Tulipa* species—tulip
- *Veltheimia* species—forest lily
- *Zephyranthes* species—zephyr lily

Kohleria species—kohleria
Sinningia hybrids—gloxinera
Sinningia speciosa—florist's gloxinia

PLANTS USED INDOORS ONLY WHILE IN FLOWER

Anthurium species—flamingo flower
Aphelandra squarrosa—zebra plant
Calceolaria species—pocketbook flower
Camellia japonica—camellia
Capsicum species—ornamental pepper
Chrysanthemum morifolium—mum
Citrus Limon—lemon
Clerodendrum species—glorybower
Crossandra infundibuliformis—firecracker flower
Cyclamen persicum—cyclamen
Euphorbia pulcherrima—poinsettia
Gardenia jasminoides—gardenia
Hibiscus species—hibiscus
Hydrangea macrophylla—big-leaf hydrangea
Kalanchoe blossfeldiana—kalanchoe
Primula species—primrose
Rhododendron species—azalea
Rosa species—miniature rose
Senecio × *hybridus*—cineraria
Solanum pseudocapsicum—Jerusalem cherry

Left: Lilies are among the flowering houseplants that need a rest period.

Above: Azalea is best used as an indoor plant only while in bloom.

If you understand the basic functions of flowering houseplants you can anticipate their needs and bring them into bloom throughout the year. Flowering houseplants need a bit more care than most foliage houseplants but when grown well they reward you with their delightful colors and scents.

Growing Flowering Plants Indoors

Growing flowering plants successfully indoors is quite different from simply decorating your home with flowering plants. Ideally, the placement of plants in a room should be a matter of taste and design. However, when the object is to grow the plant and have it produce flowers, the environment must be well suited to the plant's needs. You must provide the proper amount, duration, and quality of light; you must provide for water and fertilizer needs; and you must make sure that humidity and temperatures are adequate for plant growth.

Plants must function very well indeed if they are to carry out the process of flowering. When you grow plants indoors for flowers, you must pay more attention to the needs of the plant than when you grow plants for their foliage only. Once you understand this, you will find it worthwhile to spend the extra time required to plan growing sites carefully and select plants that are appropriate for the sites you have in mind. With a bit of practice you will be growing flowering plants in your home all year long.

The Parts of a Plant

To care for a plant properly, it is helpful to know a bit about how plants are put together. Then you will be able to anticipate the needs of your flowering plants and provide for them. Watering, fertilizing, grooming, propagating, and seasonal care—often confusing to the novice indoor gardener—are easy to carry out once you know exactly how they affect plants. Most plants are made up of four major parts: roots, stems, leaves, and flowers. Each part has its own special function, crucial to plant life, and each must work in concert with the other parts in order for the plant to grow and flower correctly.

THE ROOTS

The root system of a plant has two major functions. First, the roots serve to support the plant and anchor it in the bed or the pot where it is growing. Second, the roots are the primary means by which the plant absorbs water and mineral nutrients.

On some plants the roots have a third function—they store food, particularly during dormant periods of the growth cycle. Underground plant parts known as bulbs, tubers, corms, and rhizomes are not roots. These plant parts, commonly involved in food storage for many of our most popular flowering houseplants, are actually modified stems.

Roots absorb water and nutrients near their tips, often through tiny hairs that extend outward from their surface cells. These root hairs develop and penetrate into minute spaces among soil particles. Root tips, with or without root hairs, are extremely delicate and easily injured. Transplanting often destroys many of them. Poorly drained potting mixes, overwatering, or underwatering can also injure them. Fertilizer salts that build up as soils dry out can burn the roots and kill them.

THE STEMS

Stems transport water and mineral nutrients to the leaves, buds, and flowers through specialized vascular cells. They transport growth hormones produced at the stem tips, as well as carbohydrate food (sugars) produced by the green leaves, down to the roots and other plant parts. Stems also store food, as mentioned above. They support the plant and define the placement of leaves and flowers. Any injuries to stems can cause wilting, yellowing, and general inhibition of vigorous plant growth. On some indoor flowering plants, stems must be trained, staked, or tied to preserve proper form. It is very easy to injure a plant, and even easier to injure its flowers, during this process. Thus you should train or stake plants before the blooming process begins.

THE LEAVES

Leaves are responsible for manufacturing food through the process of photosynthesis. The foods are made up of water and carbon dioxide. The carbon dioxide (and a fraction of the water) comes from the gases of the air, and the leaves are designed to absorb and diffuse these gases efficiently. Leaves also provide a surface from which water evaporates to keep the plant cool. This evaporation of moisture creates a stream or flow of water through the plant that begins at the root, moves up the stem, and finally exits at the leaf.

THE FLOWER

Flowers contain the plant's reproductive apparatus. The colors, shapes, and scents of flowers are designed to be attractive to butterflies, bees, hummingbirds, and other pollinating creatures. A plant will initiate flowers when conditions in the natural environment are right for flowering. When you can simulate this natural environment in your home, your plants will reward your effort by producing flowers indoors.

Environments that contribute to the plant's ability to flower always provide adequate light intensity. Furthermore, this light must be of certain day length (the number of hours of light the plant receives in each 24-hour cycle) and it must also be of a certain color quality (see pages 21–22) for many plants. Since an understanding of these light requirements is so critical to your success as an indoor gardener, they will be explained in further detail later in this chapter (see pages 16–22).

Basic Plant Processes

The various parts of a plant function together, enabling the plant to carry out many life processes. The major processes are photosynthesis, respiration, transpiration, growth, and flowering. Growth and flowering depend on the success of these other three processes, which in turn depend on light, water, temperature, and fertilizer.

PHOTOSYNTHESIS

Plants produce their own food through the process of photosynthesis. The food that they manufacture is sugar, which is a carbohydrate. In photosynthesis, the energy of light is captured and used to make these sugars (which then contain light energy). Plants must have ample reserves of food and energy in order to flower properly. In fact, when photosynthesis slows down and food reserves decline, the first indication of trouble is that the plant stops blooming. Therefore, in order for plants to flower properly, they must receive plenty of light to carry out the process of photosynthesis. This is why light needs should be foremost in the minds of gardeners who wish to grow flowering plants indoors.

Photosynthesis occurs when a plant's leaves, stems, and roots are healthy and are interacting properly in a favorable environment—that is, an environment that provides correct light, temperature, and humidity. It is especially important to note that this process is a plant's only source of food and energy; no amount of fertilizer can make up for an improper environment or a lack of care that slows down photosynthesis. Maximum plant health, growth, and flowering require maximum rates of photosynthesis.

RESPIRATION

Respiration is an energy-releasing process that occurs in the cells of all living things. In respiration, oxygen is used to "burn" food, which releases the energy stored in that food. In plants, the process of respiration releases the trapped light energy that was captured in the sugars by photosynthesis. When new cells are formed, two things can happen: Either new shoots develop or the flowering process begins. Respiration provides the energy needed for each of these processes.

Respiration depends on the temperature of the air and the availability of food and oxygen. Lack of any of these can limit respiration and cause plants to languish. For some plants, temperatures are critical to respiration. African violets and other gesneriads need warm temperatures day and night for optimal respiration, while other plants require cooler temperatures or a cycle in which the temperature drops by 5° or 10° F each night. Overwatered or poorly drained soils contain very little oxygen; the roots in these soils have no oxygen to carry up into the plant, and this can prevent respiration.

TRANSPIRATION

Transpiration is the movement of water through the tissues of a plant. It begins when roots absorb water from the soil. The water moves through the roots to the stems and then into the leaves of the plant and exits the leaves by evaporating into the air. Water and nutrients are constantly being conducted through plant tissues in this manner.

Plants transpire for several reasons. Since they do not have circulatory systems, transpiration produces the vital distribution of water and nutrients throughout plants. Plant tissues are easily damaged by high temperatures and transpiration helps to maintain the temperature of these tissues, most notably the temperature of the leaves. On a sunny day, leaf temperatures are often 10° F higher than the surrounding air, even with adequate transpiration. Transpiration cools by evaporation (in much the same way that the release of sweat cools our bodies), keeping the leaf temperatures in check.

The rate of transpiration in a plant depends on the temperature and humidity of the surrounding air. The higher the air temperature and the lower the humidity, the faster plants transpire and the more water they give off. If too much direct sun or excessively dry soil causes them to transpire more water than they can absorb through their roots, they wilt.

Flower tissue is delicate. It can be thin and intricately structured. Colors are often subtle and fragile. Flowers on plants with inadequate transpiration quickly fade and wither. A plant may need some daily direct sun in order to allow enough photosynthesis to occur for flowering; but once the flowers have opened, this direct sun may damage them. Drafts can also dry out and damage flowering plants. Excessively dry air (found in many homes in the winter) often damages flowers on indoor plants. Adequate soil moisture is essential to prolong the flowers and the flowering process.

Environmental Awareness

When you plan to grow flowering plants in your home, take note of where your garden sites exist or where you can create them. In all cases, indoor flower gardens work, both functionally and aesthetically, when forethought and planning go into them.

Success with growing African violets and other indoor flowering plants involves—first and foremost—an environmental awareness on your part. If environmental needs are satisfied, plants remain healthy. In addition, they can be brought into their flowering cycle without much additional effort. Unlike most animals, plants cannot get up and move when their environment becomes inhospitable. A plant can only signal an improper environment by becoming sickly, unattractive, or by not blooming. The point is, success with plants comes with the awareness that allows you to create the right environment or move the plant to a new environment before it suffers. Obtaining this awareness means that you recognize where in your home plants will do well. Also, you must learn to recognize how a certain indoor

Commercial growers perfect large-scale growing environments. You can succeed on a smaller scale by carefully choosing and modifying indoor gardening sites.

gardening site can be created or improved with lights, air movement, and temperature control. All of this should be carried out before plants are placed in a new location.

Providing Light

The biggest challenge you will face when you attempt to bring plants into bloom indoors will be to provide them with enough light. With any given plant, flowering generally requires more light than is needed for simple maintenance of the plant's foliage. Most indoor flowering plant gardeners greatly increase their success as they realize this light intensity need. But the challenge goes beyond providing the quantity of light needed; the light must also be of the right quality and of proper duration (day length). You will be able to prevent or solve plant problems resulting from improper light if you are equipped with an understanding of what light is and how it behaves.

Plants need light for several reasons. Certain kinds of light influence the production of essential hormones, many of which help to initiate the flowering process. Still, the most essential reason a plant needs light is that photosynthesis will not occur without it. Plants photosynthesize food in direct proportion to the amount of light they receive (as long as the light is not at heat-damaging levels). And well-fed plants are generally healthier, more vigorous, and far more likely to flower.

LIGHT INTENSITY

Most information written about the culture of African violets devotes few paragraphs to light intensity. This is probably due to the fact that African violets flower in significantly less light than almost any other kind of flowering indoor plant. Some of the bulbous plants that can be forced into flower indoors also respond to moderate light intensity. But many other types of indoor flowering plants require higher light intensities to bloom.

Bromeliads, especially those with gray-green leaves, tolerate a full-sun setting.

Whereas indoor gardening can be done with plants in sites varying from "full sun" (four or more hours of direct sunlight per day) to those in "shade" (lighted spots that are comfortable for reading), indoor gardening of flowering plants is generally limited to those sites varying from the moderate light in African violet stations to "full sun" locations. As the plant encyclopedia points out, many flowering "house-plants" are best suited for greenhouse sites—at least until they have been brought near the midpoint of the flowering process. At that time, they can be taken to more moderately lit presentation sites in the home.

Light Requirements for Flowering Indoor Plants

PLANTS FOR DIRECT SUN

Amarcrinum memoria-corsii—crinodonna
All Bromeliads
Capsicum species—ornamental pepper
Citrus limon—lemon
Crossandra infundibuliformis—firecracker flower
Gardenia jasminoides—gardenia
Hibiscus species—hibiscus
Hydrangea macrophylla—hydrangea
Lachenalia species—Cape cowslip
Lilium species—hybrid lilies
Pelargonium species—geranium
Primula species—primrose
Rosa species—miniature rose
Sinningia speciosa—florist's gloxinia
Solanum pseudocapsicum—Jerusalem cherry
Sprekelia formosissima—Aztec lily
Tulbaghia fragrans—society garlic
Zephyranthes species—zephyr lily

Orchids prefer sunlight that has been softened by a sheer curtain.

PLANTS FOR CURTAIN-FILTERED SUN

Abutilon species — flowering maple
Achimenes species — rainbow flower
Aeschynanthus species — lipstick plant
Agapanthus species — lily-of-the-Nile
Anthurium species — flamingo flower
Aphelandra squarrosa — zebra plant
Begonia × *cheimantha* — Christmas begonia
Begonia × *hiemalis* — Rieger begonia
Begonia × *semperflorens-cultorum* — wax begonia
Begonia × *tuberhybrida* — tuberous begonia
Camellia japonica — camellia
Clerodendrum species — glorybower
Clivia miniata — kaffir lily
Columnea species — columnea
Crinum species — Bengal lily
Crossandra infundibuliformis — firecracker flower
Daphne odora — winter daphne
Episcia species — flame violet
Exacum affine — Arabian violet
Gardenia jasminoides — gardenia
Haemanthus species — blood lily
Hibiscus species — hibiscus
Hyacinthus orientalis — hyacinth
Impatiens wallerana — Patient Lucy
Kohleria species — kohleria
Nematanthus species — nematanthus
Orchids

Amaryllis does well in bright, indirect light — about 5 feet back or a bit to the side of a window that gets some sun.

Ornithogalum species — star-of-Bethlehem
Rosa species — miniature rose
Scilla species — squill
Senecio × *hybridus* — cineraria
Sinningia hybrids — gloxinera
Sinningia speciosa — florist's gloxinia
Streptocarpus species — Cape primrose
Veltheimia species — forest lily
Tulipa species — tulip

Below: Narcissus is among the plants that do not require a particular level of light once they are blooming.

PLANTS FOR BRIGHT INDIRECT LIGHT

Chrysanthemum × *morifolium* — mum
Crocus species — crocus
Episcia species — flame violet
Eucharis grandiflora — Amazon lily
Euphorbia pulcherrima — poinsettia
Hippeastrum hybrids and species — amaryllis
Hoya species — wax plant
Lilium longiflorum — Easter lily
Muscari species — grape hyacinth
Saintpaulia species — African violet
Schlumbergera species — Christmas cactus
Spathiphyllum species — peace lily

PLANTS THAT CAN BE PLACED ANYWHERE WHILE IN FLOWER

Calceolaria species — pocketbook flower
Cyclamen persicum — florist's cyclamen
Kalanchoe blossfeldiana — kalanchoe
Narcissus hybrids and cultivars — daffodil
Primula species — primrose
Rhododendron species — azalea
Tulipa species — tulip

Measuring Light in Footcandles

A footcandle is the unit of light intensity measured by light meters. Knowing the number of footcandles of light present at various plant stations in your home will help you predict how well plants will perform there. As you become more familiar with indoor gardening, you will be able to judge light intensity in less precise terms. In our plant encyclopedia we have not given footcandle light requirements, but have expressed light intensity requirements as "direct sun," "curtain-filtered sunlight," and "bright indirect light." As the following chart illustrates, light intensities vary markedly in any home.

This chart is a good starting point, but the only objective way for you to determine the light intensity of an indoor growing station is to use some sort of a light meter. The best type of light meter is an incident (incoming) light meter, which reads directly in footcandles. Professional interior landscapers generally have these meters, which cost from $50 to $200. With reasonable success, you can use the reflectant light meters commonly found in cameras that have through-the-lens light metering. Since the reading from the camera is not in footcandles, you will have to use the conversion method given in the box at right.

A camera with a built-in meter offers a simple way to read light in footcandles.

Environmental and Geographical Effects on Light

You may have noted that the intensity of light varies over the year, even within one particular plant station. Seasonal changes in the angle of the sun or the amount of shading from outside, day-to-day weather or atmospheric changes, and the varying presence of nearby reflective surfaces make quite a difference to your plant's well-being. Failure to react to such changes in light intensity often results in plants that do not flower.

Summer sun shines almost perpendicular to the earth, striking it with maximum intensity. At any window on a rainy day in winter, the light intensity may be only one twentieth of that in the summer. Even the most tender begonia that could never withstand full summer sun will welcome full winter sun. Of course, southeast windows with no obstruction may need a sheer curtain even in winter to protect the flowers from sunburn. You should also keep in mind that winter sun reaches further into south-facing rooms because of the low angle of the sun. Plants may get a few hours of full sun in the middle of a room in winter; but in the warmer seasons the situation changes and you may need to move these plants closer to a window in order to give them adequate light.

Where you live also affects how much light you receive. For example, sunlight in the Rockies during winter is much more intense than in New England, because higher elevation means thinner air and less light diffusion through the atmosphere. In winter, the sun rises and sets farther to the south of the United States. Consequently, Florida and other southern states receive more bright light than northern states such as Montana.

Light intensity will vary even within your local area, and not just because of the seasons. Industrial smoke may make sunny days hazy, and clouds or fog will cut down light. Trees and shrubs that shade your home reduce the amount of light that passes through

Footcandles of Light Generally Present in Several Common Sites*

0 500 1000 2000 3000 4000 5000 6000 7000

OPEN SHADE OUTDOORS IN SUMMER—4,000 TO 8,
CLOUDY WINTER DAY OUTDOORS—500 TO 2,000
SOUTH-FACING WINDOWS, CURTAIN-FILTERED SUN—1,000 TO 3,000
BRIGHT, INDIRECT WINDOW LIGHT—600 TO 1,000
NORTH-FACING WINDOWS, SUNNY SUMMER DAY—2,000 TO 4,000
NORTH-FACING WINDOWS, SUNNY WINTER DAY—500 TO 1,000
FLUORESCENT FIXTURE, TWO 40-WATT BULBS (6 INCHES BELOW)—600 TO 800
FLUORESCENT FIXTURE, TWO 40-WATT BULBS (12 INCHES BELOW)—200 TO 500
COMFORTABLE READING LIGHT—50 TO 100
HALLWAYS—10 TO 50

your windows. The screens on windows, doors, or porches reduce light by at least 30 percent. A white house next door or a light-colored cement driveway will reflect sunlight, increasing the intensity of light that your rooms receive. Snow also reflects a great deal of light, especially on a sunny day.

Using a Camera's Light Meter to Determine Light Intensity in Footcandles

Since the camera's meter is measuring the amount of light reflected from a surface, the first thing you must do is place a fully reflective (white) surface in the spot where the plants will be. This can be done with a white piece of paper or poster board. Place the camera on the same side of the paper as the light source, but do not cast a shadow onto it. This may be difficult to do in a window box, but it is necessary if the right amount of light is to be detected and measured. Fill the camera's viewfinder with the image of the white surface. Set the film speed setting to ASA 50 and put the shutter speed on 1/50th of a second. Adjust the aperture setting (F-stop) until the exposure reading in the camera indicates "correct" exposure. The approximate light intensity in footcandles is indicated in the following table:

F-Stop Setting	Footcandles
4	300
5.6	600
8	1200
11	2400
16	4800
22	9600

LIGHT DURATION

Most plants are adapted to thrive and flower when given specific day-night cycles or day lengths. Many plants, such as poinsettias, will not flower at all unless they are given uninterrupted nights, lasting for at least 12 hours, for two or three weeks. If the plant is near a light in the house that is turned on during these 12 hours, the plant will not flower. Other plants, such as Reiger begonia, kalanchoe, or Christmas cactus, do not require such treatment; they flower more satisfactorily, however, if they are provided with long uninterrupted nights for a period of time. Still other plants, like African violets, are called "day-neutral." They flower under any length of day or night cycling.

Even day-neutral plants need alternating periods of dark and light, however. The number of hours of light required depends to some extent on the intensity of the light, although the two are not in direct proportion. Many plants that will not bloom because the light intensity is too low can be induced to flower by increasing the number of hours of "daylight" they receive. This explains why some day-neutral indoor plants will flower in the summer, even though they are shaded because of the higher angle of the sun or by the foliage of trees outdoors; the day length is longer even though the light they receive is not as intense.

Actually, most listed light requirements for indoor plants are based on a 12-hour day length. Although supplemental electric lighting may be needed for adequate flowering in the winter, 16 hours of light in each 24-hour period is the longest day length you should ever give your plants. After this, light becomes damaging to the plant.

Fluorescent lamps pictured are (top to bottom) daylight, deluxe warm white, warm white, cool white, Gro-Lux, Agro-Lite, and Vita-Lite.

LIGHT QUALITY

The sun emits a continuous spectrum of colored light from ultraviolet, through the visible colors, to infrared. All these colors of light together create white light. The colors are usually combined in sunlight, but they are separated in sunrises, sunsets, and rainbows. The color mix of a light source defines the "quality" of that light.

Plants respond more favorably to certain colors in sunlight than they do to others. The infrared color that makes up about 50 percent of the energy of the sun's light can be harmful to plants. When it is absorbed by leaves, it rapidly turns to heat energy and can lead to sunburn. On the other hand, some infrared (and red) light is

9000 10000 11000 12000 13000 14000 15000

SUNNY OUTDOOR SPOT IN SUMMER—10,000 TO 15,000

UNNY SPOT OUTDOORS IN WINTER—7,000 TO 10,000

OUTH-FACING WINDOWS IN FULL SUMMER SUN—8,000 TO 10,000

** The footcandle ranges result from measurements of average situations at about 42 degrees North latitude, in the center of unobstructed windows.*

needed by plants to influence the hormones that ultimately permit the flowering process to begin.

Electrically produced ("artificial") lights vary in quality. Moreover, they do not produce a continuous spectrum of colors in the same manner as sunlight. For instance, incandescent lights produce mostly red light. Fluorescent lights vary greatly, but most emit light largely in the blue region. The fluorescent "plant" or "grow" lights are designed to produce blue and red light closer to the mix needed by plants.

When you measure light with a light meter, you are measuring whatever color light is present. Thus footcandle measurements can be misleading if the quality of light is not taken into consideration. You may not have as much useful light as you think you have. In the following table, the "plant usefulness" of light is given relative to daylight (which equals 1.0 in the table). Remember, those typical footcandle measurements given on page refer to the intensity of sunlight. The measured footcandle intensity of cool white or warm white fluorescent lights will have to be almost doubled to equal the light needs given in a plant list.

Designers have long used lights of varying quality to influence the color rendering of plants, people, and furnishings. This can be an important consideration when growing African violets and other flowering indoor plants. The reddish-looking grow lights enhance the color of plants with blue, yellow, or red flowers. When lights over plant stations must also serve to light furniture and people, grow lights can be distracting. In such cases, most indoor gardeners compromise by using a combination of cool white fluorescent, warm white fluorescent, and a small amount of incandescent light.

Watering Techniques

The quantity of water a plant needs depends on the conditions under which it is grown as well as the genetic inheritance or natural background of the plant. For instance, African violets are jungle plants from an area where rainfall is high and fairly evenly distributed throughout the year. This makes them poorly equipped to deal with drought. Though prolonged underwatering may not kill them, it will certainly weaken them and should be avoided. They grow best when watered thoroughly, and then allowed to dry slightly between waterings. These genetic needs of the various indoor flowering plants are given in the encyclopedia section of the book.

The phrase one hears most in connection with watering houseplants is "keep the soil evenly moist." This can be misleading. It does not mean to water the plants a little bit every day, nor does it mean the soil cannot dry out at all. It means that the soil should be watered thoroughly when it is slightly moist, and then allowed to dry to the point of being only slightly moist again before being watered.

Quality of Light Sources Commonly Used for Plants

LAMP	PLANT USEFULNESS FACTOR	COLOR IMPROVED	COLOR GRAYED
Fluorescent			
Daylight	0.53	Green, blue	Red, orange
Cool White	0.53	Blue, yellow, orange	Red
Warm White	0.52	Yellow, orange	Blue, green, red
Deluxe Warm White	0.52	Red, orange, yellow, green	Blue
Gro-Lux	1.15	Blue, red	Green, yellow
Gro-Lux Wide Spectrum	0.78	Blue, yellow, red	Green
Agro-Lite	0.72	Blue, yellow, red	Green
Vita-Lite	0.68	Blue, yellow, red	Green
Incandescent	1.55	Yellow, orange, red	Blue

** The plant usefulness factor is derived by computing the amount of radiation emitted in the 400–850 nm (nanometer) wavelength range (the wavelength of useful light for plants) and comparing it to daylight (which is set to 1.0). The figure is misleading with incandescent light, since very little light is emitted in the 400–500 nm (blue to green) part of that range!*

Make sure you water thoroughly and drain off the excess water that collects in the saucer beneath the plant. The need to water adequately and let the soil dry to the right level between waterings cannot be stressed too much. Failure to understand this leads many people to feel that they have no luck with plants, especially flowering plants.

If you are just starting a collection, or have recently purchased plants, it is a good idea to check them every day until you learn how often they need water. The water needs of your plants will vary through the year. Light intensity, day length, temperature, and the humidity in the air have a strong effect on water use. Sometimes plants need water every three or four days, and at other times once a week is enough. Never water plants strictly on the basis of the calendar.

Most indoor flowering plants, especially African violets and other gesneriads, should be given water that is room temperature or slightly warmer—what is commonly referred to as "tepid" water. Cold water is a shock to many plants and can be harmful.

Artificially softened water is dangerous to the health of your plants. If you have a water softener in your home, take the water for your houseplants from an outdoor faucet or, if possible, from the pipe indoors before it has been through the softener. The minerals in hard water are not nearly as bad for your plants as the sodium that the softener adds.

Every now and then your plants may go dormant or "rest." African violets vary in this regard. Some seem to bloom without interruption, but others occasionally rest for a few weeks. Other indoor flowering plants, such as gloxinias and many bulbous plants, require a complete dormancy period between flowering cycles. Our plant encyclopedia is organized according to such dormancy needs.

Generally, plants go dormant after a period of heavy bloom, when the weather is especially dark and cloudy, or during very hot or cold periods. They are most susceptible to overwatering when entering this resting or dormancy cycle, so check them carefully and water only when they need it. Get into the habit of checking your plants as often as you can. Even if they do not need water they benefit from the attention, and you will find that their care will soon become second nature to you.

OVERHEAD WATERING

There are two ways to water indoor flowering plants: overhead or top watering and bottom watering. Both methods have their pros and cons. Most gardeners water their plants from overhead. Using a watering can, apply water slowly to the surface of the soil until the soil is thoroughly wet and some water drains into the saucer beneath. Empty the excess water from the saucer as soon as the soil has drained thoroughly.

One drawback to overhead watering is that it is very easy to splash water on the foliage. African violets are particularly susceptible to leaf-spotting, which is caused by cold water splashed on the leaves. The cold water causes a sudden temperature change and, in the presence of strong light, this kills some of the surface cells of the leaf. In the Gesneriad family, most of the chlorophyll in the leaf is in a single layer of surface cells. When this layer is killed, the chlorophyll fades and the underlying leaf color—usually a greenish white—shows through. The water drops often make a spot on the leaf in the shape of a ring or arc. These spots do not seriously injure the plant, but they are unsightly.

Leaf-spotting can be prevented by watering carefully so that the leaves don't get wet, and by using tepid water. Never splash water onto the leaves of African violets unless you are cleaning them. Some people prefer to water their plants from the bottom to avoid the risk of leaf-spotting.

BOTTOM WATERING

Many people find that watering plants from the bottom through the drainage holes is easier and faster than top watering. Make sure the saucers you have the plants in will hold enough water to saturate the soil. At watering time, pour water into the saucer. Capillary action will cause the water to seep up and moisten the entire rootball. One hour after watering, drain off any excess water that is left in the saucer. Plants that are watered from below must be leached or flushed out at least once every few months. Leaching washes out mineral salts that can build up in the soil and damage the plant. To leach a plant, water it copiously from above, let it drain, then repeat the process two or three more times. This is most easily done by moving the plants to a sink, to a bathtub, or outdoors.

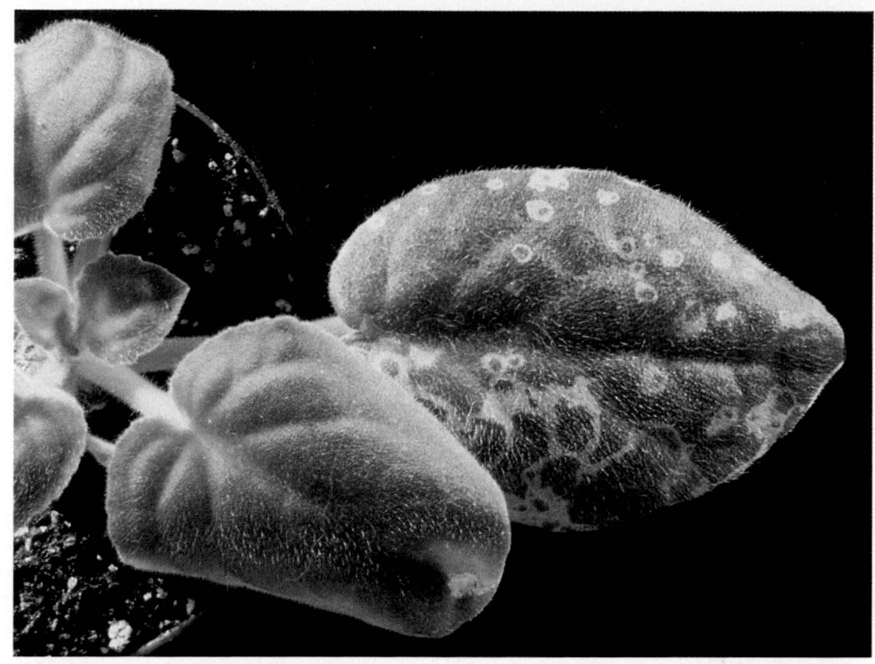

Water spots on African violet leaves are caused by splashes of cold water.

To protect delicate foliage, plants can be watered from below.

A turkey baster provides an easy way to remove excess water.

In this homemade wick watering setup, a frayed nylon wick carries water from the saucer up into the soil.

VACATION WATERING

Before you leave on vacation, you need to be sure that your plants will be able to survive without your constant care. Water all plants thoroughly and allow them to drain. Place them in bright light, but do not leave any in direct sun. Remove dead or dying leaves, faded flowers, and flower buds that will open when you are away. Finally, check for any insects or disease problems that should be dealt with before you leave. The pictured watering devices will keep your plants healthy for several weeks, and some can even be used regularly for routine watering.

Fertilizing Techniques

The fertilizers you buy to make your plants grow well are often called "plant food." This can be misleading. Fertilizers are not food. They are complex products that contain the nutritive substances plants need to develop tissues and keep cells functioning. We have similar "fertilizer" needs, but we call these elements in our diet vitamins and minerals.

When fertilizing African violets and other indoor flowering plants, stay away from products that are too high in nitrogen, such as those with a 30-10-10 or 2-1-1 NPK (nitrogen-phosphorous-potassium) ratio on the label. These promote lush foliage growth at the expense of bud production. The formulas most often recommended for flowering plants are 15-30-15, 10-10-5, and 5-5-10. Many growers use two formulas—an all-purpose fertilizer such as 20-20-20 or 18-18-18 and a "bloom food" such as 0-10-10. A bloom food should be used only for an extra boost when flower buds become visible on the plant; do not use it to attempt to force the plant to bloom.

Many indoor gardeners have found that they do not need to apply fertilizers as often as the label directs. This is because plants do not grow at a constant rate throughout the year. Also, a number of fertilizers are formulated for use with fast-growing foliage plants in sites with high light intensity. African violets and other gesneriads, begonias, and some bulbous plants do not require as much fertilizer as other plants require. A good guideline is to apply fertilizer half as often as recommended on the label, or to dilute to one half the strength recommended; some do both. You will find that the plants perform just as

Three purchased self-watering devices:

Top: Porous clay pulls water through tubing to keep soil moist. One water source can be concealed behind several plants.

Above: A double-walled pot provides an ample and easily refilled reservoir.

Right: Absorptive felt serves a wick to pull water from the saucer reservoir into the soil.

well, and you will be saving both time and money.

With most flowering indoor plants, it is also a good idea not to fertilize all the time. Many bulbous plants should not be fertilized until they have ceased flowering. Most plants should be fertilized less when they are not actively growing.

African violets can be fertilized on a "constant-feed" program, in which the plants are fertilized every time they are watered, all year around. When they are fertilized this way, plant growth is more symmetrical and leaf color and size is more even. You must use a fertilizer designed especially for this purpose, however. If you do not, you risk overfertilizing and damaging the roots. Plants that require less watering during rest periods also require less fertilizer during these times.

An important point to remember: Never give fertilizer to a sick plant unless you are absolutely sure the plant is suffering from a lack of nutrients. When plants are ailing, they cannot use the fertilizer and you run the risk of burning the roots because the salts from the fertilizers build up rapidly.

Atmospheric Needs

The right atmosphere—a combination of temperature, humidity, and air circulation—is very important to plants. Many indoor gardeners have trouble because they are not aware of these atmospheric needs. In addition, many are not aware of the variations and changes in indoor atmospheres from site to site within the home or from season to season throughout the year.

TEMPERATURE

Although many indoor flowering plants prefer cool temperatures (especially at night), African violets should never be kept too cool at night. Days of 60° F to 70° F and nights of 65° F to 70° F are best. They also benefit from being kept free from drafts and from rapid temperature drops at the end of the day.

Many of the flowering plants in the encyclopedia, such as episcias, begonias, and many orchids, are tropical plants that are native to areas where temperatures and humidity are high. These do best near a south-facing window in a room with an appliance that vents moist heat, such as a dishwasher, a clothes dryer, or a humidifier. Cool-loving plants (55° F to 60° F days and 50° F nights), such as cyclamen, camellias, azaleas, and some orchids, do well in rooms where indirect sun keeps temperatures low.

HUMIDITY

Because of the general fragility of flower tissue, nearly all flowering houseplants, including African violets, prefer a humidity level of 50 percent or higher. In drier climates, or in the winter, it is practically impossible to create this level of humidity in a home.

A cool-vapor humidifier is one excellent way to increase humidity. Place portable units wherever they are needed. You may even want to install a humidifier as a part of your home's central heating system.

The simplest method of humidifying the air around flowering plants is to set the pots in trays or saucers filled with pebbles, perlite, or vermiculite. Fill these with enough water to reach just below the surface of the material. Be careful not to add so much that the bottom of the pot touches the water; this can cause the roots to rot.

To examine the roots of a large plant, grasp the pot, placing your fingers over the edges of both pot and rootball. Tap the edge of the inverted pot on a solid surface, then ease the rootball out, using your thumbs to lift back the pot.

Examination shows that this chrysanthemum has become badly root-bound.

Water in pebble trays should never be deeper than the tops of the pebbles.

Soil that is kept too soggy can cause healthy roots (left) to decay (right).

AIR CIRCULATION

Drafts of dry air cause stress and can easily burn the tissues. Be careful if you keep plants near a window during winter — cold drafts and frosted window panes can harm them. Hot-air registers are a source of drafts of dry air. Fumes from burning propane or butane gas may damage flowering plants and cause them to drop their buds.

Potting and Repotting

The roots of plants grown in containers are confined. With most plants, including African violets, this confinement generally promotes flowering. But after a while either the roots get too crowded, the plants get too big, salts build up in the soil, or the soil mix decomposes and is no longer aerated. These conditions commonly cause plants to stop flowering and you must repot them.

It is a good idea to get into the habit of knocking plants out of their pots and examining the roots at least once a year. Root problems often go unnoticed until too late. Gently slip your fingers under the leaves and cradle the stem between your third and fourth fingers. Invert the pot and lightly rap on the sides and bottom until the rootball slides out. Examine the roots. If there are so many roots that the outside of the soil is matted and white with them, or if the roots are brown and rotted, it is time to repot the plant — and the sooner the better.

Most nurseries carry (left to right) plastic pots; shallow, unglazed clay fern pots; glazed ceramic pots; and the traditional tall, unglazed clay pots.

An easy way to mix soil is by lifting alternate corners of a plastic tarp.

THE RIGHT CONTAINER

The only hard-and-fast rule in choosing a container is to always use a pot with a drainage hole. Unglazed clay pots are traditional, but plants can thrive in containers of any material. Shallow pots (called fern or azalea pots) are good for the shallow fibrous root systems of African violets. Unglazed clay pots dry out quickly, and fertilizer salts pass out to the surface of a clay pot rather than building up in the soil. Because of this you may not have to leach the soil as often, but wherever the stems of African violet leaves touch the rim of the pot, the stems will burn and leaves will be lost. To prevent this, coat the rim of the pot with wax.

No matter what kind of pot you choose, make sure the size is appropriate for the plant. If a small plant is put into a pot that is too large, the excess soil stays wet and the roots are likely to rot. If your plant's roots are rotted or unhealthy, repot the plant, using fresh soil, into a pot of the same size or even a smaller pot, depending on the condition and size of the roots.

THE RIGHT SOIL

Many commercial indoor soil mixes are too fine and poorly aerated for the fibrous root systems of indoor flowering plants. Bulbous plants, on the other hand, will tolerate almost any kind of potting mix. In a good potting mix there should be obvious particles of leaf mold or bark in the mix; easily seen fibers of peat moss; as well as some coarse sand, vermiculite, or perlite. When you open the bag, reach in and grasp a handful of the mix. You should not be able to form it into a ball in your hand. If you can mold it as you mold clay or loamy soils, it is too heavy.

THE RIGHT TIME

Avoid potting or repotting plants while they are in flower. No matter how careful you are, fine roots will be damaged and stress will be put on the plant. Although stress is not usually harmful to foliage plants, in flowering plants it can often lead to wilted flowers and can bring a prompt end to the flowering process. Since African violets bloom steadily given the right care and conditions, it is a good idea to remove all the buds and blossoms from a plant when you repot it, so the plant can recover before it flowers again. African violets need to be slightly root-bound in order to bloom properly, so pinch off buds to prevent flowering for several weeks after you repot them. Plants that are purchased already in flower can be slipped into another pot for decorating purposes and then repotted when they have finished blooming. Finally, it is not advisable to repot during a hot spell. Fall or spring are the ideal seasons for replanting most flowering plants.

Creating African Violet Hybrids

The process of crossing two African violet plants to make a hybrid plant is not as difficult as it may seem. From start to finish, it takes two to three years before you see your new hybrid flowers, but knowing that you created the hybrid yourself more than compensates for the wait.

1. African violets are perfect flowers, which means that they have both male and female organs in the same flower. The female organ is called the pistil and consists of the stigma, the style, and the ovary. The male organ is called the stamen and consists of the filament and the anther. The anther contains the pollen.

2. You must first choose two African violets that you want to hybridize. Make one of the plants the female parent by removing the anthers from a flower just after it opens. Do this with a small pair of scissors. You are going to put the pollen of the second plant on this plant's stigma, so you must wait several days until the stigma is sticky.

3. Using a pair of tweezers, remove an anther from the second plant, your male parent. To check that the pollen is ripe, break open this anther and make sure the pollen shakes out easily. Take another anther, break it open, and rub it lightly on the stigma of the female plant. The pollen that sticks to the stigma will move down into the pistil and fertilize the plant. Label the plant with the date and the names of the two parent plants.

If all went well, the seedpod will begin to swell in four to seven days. It takes six to nine months for the seedpod to mature completely. When it is brown and shriveled the seeds inside should be ripe. Remove the flower stalk with the seedpod attached. Sow the seeds immediately, or keep them in a warm, dry place (they can be stored for as long as six months).

Propagating Indoor Flowering Plants

Most indoor flowering plants are produced from stem cuttings. Bulbous plants are propagated from divisions or small bulblets or offshoots. African violets are one of the few plants that can be propagated from single leaves. Specific propagation methods are given for each plant in the gallery section of our book. The following discussion will familiarize you with common propagation techniques.

DIVISION

This technique involves dividing one entire plant, including its root system and foliage, into two or more separate plants. Plants that form new plant clusters at their base (multiple basal stems) are suitable for propagation by division. These include wax begonias, most bromeliads, and cluster-forming succulents.

Two yellow stamens remain by the fuzzy African violet seedpod.

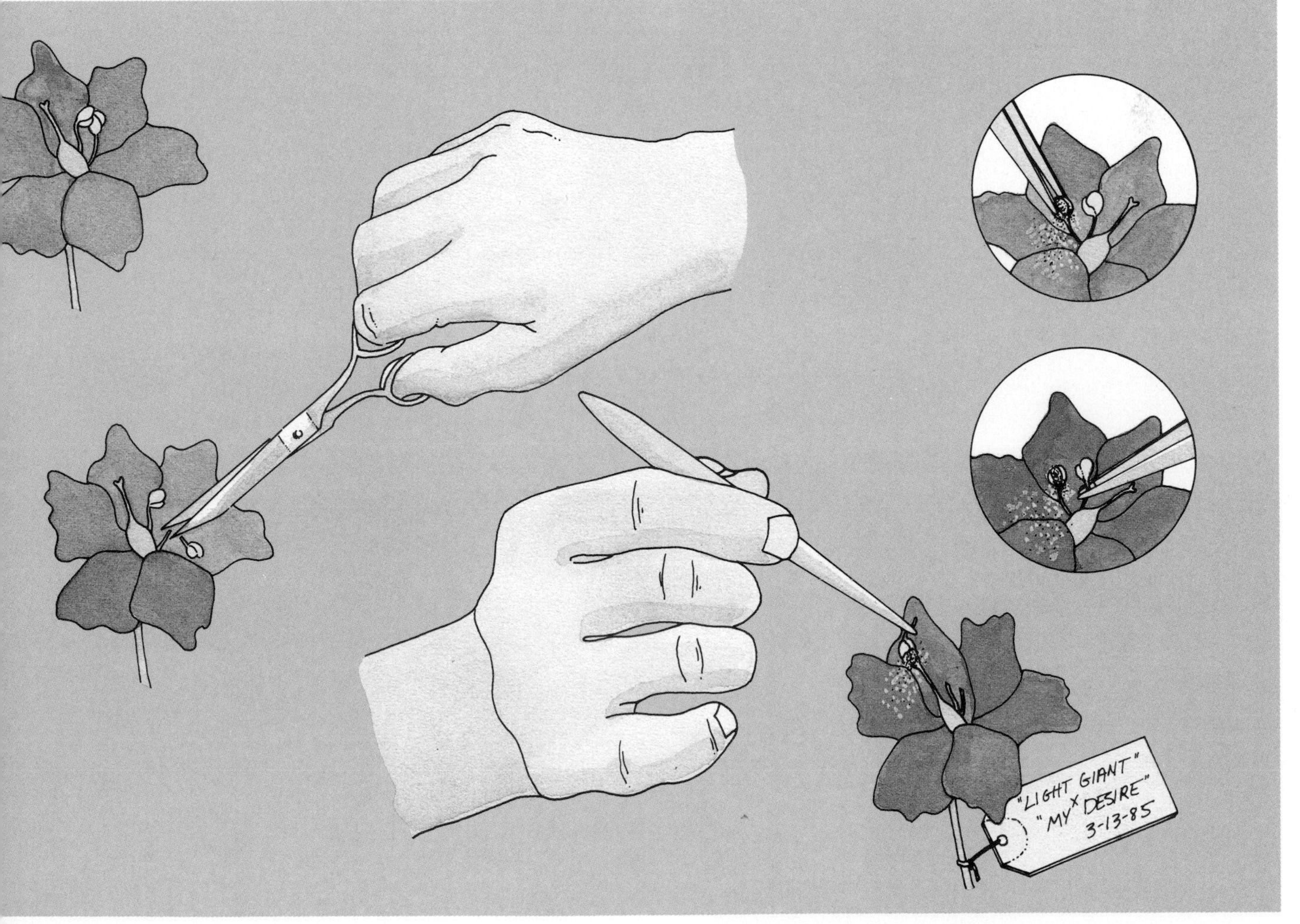

Above: Bromeliads form offsets while blooming or up to several months later. When they are firm and mature they can be removed. They will root and grow if potted and given the same conditions as the parent plant.

Below: Soft-stemmed plants, like geranium, impatiens, and chrysanthemum, root quickly from stem cuttings. Remove cutting from plant with a clean cut, trim off lower leaves and any flowers. Dip bottom of stem in rooting hormone and bury up to remaining leaves in a cutting mix, in a hole made with a pencil. Label and water. Rooting should occur in 3 to 8 weeks. Keep cutting covered with a clear plastic cup or plastic tent and out of direct sun at first, then harden off gradually.

STEM CUTTINGS

Inducing a stem to form roots is the most popular method of vegetative propagation. Cut a piece of branch, remove the lower leaves, plant the stem in a small pot, and roots will form along the buried stem. Once several branched roots have formed, it is ready to be planted in a larger pot. Most plants with soft stems, such as impatiens, are easy to propagate by this method. Others with woody stems, such as the older growth on camellias, are more difficult.

LEAF CUTTINGS

African violets are commonly propagated by leaf cuttings. Other flowering indoor plants propagated from leaf cuttings include some begonias, streptocarpus, kalanchoes, and gloxinias. To root leaves, start with any mature leaf from the middle of the plant. Do not use an obviously old leaf. After removing it from the mother plant, cut the leaf stem (petiole) to from ½ to 1 inch in length. Cut away the top two thirds of the leaf if the leaf is rather large.

The leaf stem can be rooted in water or soil, although rooting in water is the most common method. If you are rooting the plant in water, put the stem in a small jar filled with water up to the base of the leaf and keep the water level constant. Remove the leaf from the water when the roots are clearly visible, but not longer than 1 inch.

Rooting the leaf cutting in a potting medium is actually easier in the long run. Set the stem as shallowly as possible on an angle (about ½ inch deep) in the rooting medium. The little plantlets grow at the base of the leaf and are ready to divide and pot up in two or more months. When the plantlets are about 1 to 2 inches high, remove them from the pot, separating them and giving each its own pot. The mother leaf can be thrown away. If the plantlets are not divided when they are small, the plant will grow into a multiple crown "bush" with poor symmetry and only a few blooms.

Leaf cuttings are commonly used to propagate African violets. Cut a leaf that is mature but in good condition. To root it in soil, shorten the leaf stem to between ½ and 1 inch and insert it in moistened cutting mix. Cover with plastic for the first few days. Several small plants will form and be ready to pot in two or more months.

GROWING PLANTS FROM SEED

African violets and other gesneriads, impatiens, and primroses are easily raised from seed. They can be placed on a damp medium, under lights, and will grow fairly quickly if you follow the instructions in the box on the next page. They should be kept moist, never wet, and potted into larger pots as they mature. Many other flowering houseplants require a lot more time, patience, and skill. Difficult-to-start seeds often require bottom heat and overhead misting for success.

Above: African violet leaf stems root easily in water. They can be potted in soil when roots are 1 inch long.

Below: A lit growing station allows rooting of dozens of African violet leaf cuttings.

Check plants that should have a long blooming period occasionally for new buds—a sign that all is well.

Controlling and Preventing Pests, Diseases, and Other Problems

Plants must be healthy and vigorous in order to flower properly. Many plant sicknesses result from environmental imbalances (stresses) rather than from diseases or pests. Moreover, plants weakened by stresses are more easily attacked by microbes or pests. Managing stress is the first step toward keeping your houseplants healthy.

Reread the chapters on culture from time to time and review your routine. Are you watering properly and feeding enough? Are plants crowded? Do you repot them often enough? Has the light intensity or temperature changed? Attention to these details keeps problems to a minimum. Remember that no aspect of care exists apart from any of the others. They are all connected in some way. Temperature affects water needs, light intensity affects temperature, and so on.

Inspect each plant in your collection closely every few months. Examine the main stem, the undersides of the leaves, and the soil surface for insects or problems that are in their beginning stages. Knock small plants out of their pots and examine the roots and soil.

RELUCTANT BLOOMERS

At some point you have probably brought home a beautifully flowering plant, only to watch it finish blooming and never bloom again. There are many possible reasons for this, but the usual cause is insufficient light.

Try giving the plant progressively more light. If it is not close to a light source, move it closer to the window or the lights. A drastic change in light intensity may burn the foliage, so make the change gradually. But don't be alarmed if the move bleaches out the foliage a little. Unless the leaves turn yellow, the plant will not be damaged. Placing a plant closer to a light source will cause it to dry out faster. Keep a careful eye on the plant so you can adjust to its new watering needs.

If a plant will not bloom under good light, it may need a change in day length. If the lights are usually on for 12 hours, try a day length of 10 hours for a few weeks, then go back to 12 hours. If this does not work, try a few weeks at 14 hours of light a day (watch carefully to be sure this change does not create too much water stress).

Recently transplanted plants are not likely to flower for a time, as discussed earlier on page 28. Transplanting disrupts the plant's normal growing routine and the roots need time to grow into the new soil. Many plants are simply too young to produce flowers. Often, woody plants must grow for a year or two before they will flower; and bulbous plants started from small bulbs will not flower for two or three years. Most of the newer cultivars of African violets begin flowering when they are quite young, but this is not often the case for other plants.

Pruning and shaping enhances flower production, but don't get carried away with the job and prune so much or so often that buds do not have a chance to form. Many plants also need to be pinched back after blossoming. Specifics on pruning to enhance flowering are given in the plant encyclopedia.

It is extremely important to prune the outer leaves of African violets in order to maintain their proper blooming habit. Rows of outside leaves have to be removed to give the inside rows a chance to produce flowers. Each row of leaves will produce flowers one time only. The third row of leaves, counting from the center outward, carries the mature blossoms. The new flower buds are produced in the center of the plant, at the first row of leaves. Once a plant has more than five rows of leaves, the blossom number and size will gradually decline until the plant eventually stops flowering altogether.

Perhaps your reluctant bloomer just needs a rest. You can try withholding fertilizer and slightly reducing water for a month, then resume your regular care. The point is that sometimes flowering plants get into a rut and need a little contrived change of seasons or conditions to force them to develop flower buds. Experiment. You have nothing to lose, and you will learn from your experimentation. If everything you try fails, perhaps it's time to make room for a new plant in your collection.

Propagating Flowering Houseplants From Seed

1. Use a clean, 3-inch-deep container. Fill with sterilized, fast-draining, light planting medium. Drench the container and soil with a fungicide solution such as benomyl or captan. Allow the soil to drain.

2. If the seeds are tiny and dustlike, as are African violet seeds, be sure to sow them in an area that is free from drafts.

3. Label the rows of seed. Cover the container with clear plastic or glass and place in a warm, bright place out of direct sun. The temperature should be 75° F to 85° F. If moisture accumulates on the cover, wipe it dry. African violet seeds germinate erratically, and may take anywhere from two weeks to several months.

4. Remove the cover after the seeds germinate, and keep them in bright light. If grown under artificial lights, place the containers on blocks so that the seedlings are 3 to 4 inches below the lights. Keep the soil damp but not wet, and never let it dry out. Feed the plants once a week with a one-quarter-strength all-purpose fertilizer. When seedlings are 1 to 1½ inches high, transplant each to a 2¼-inch pot.

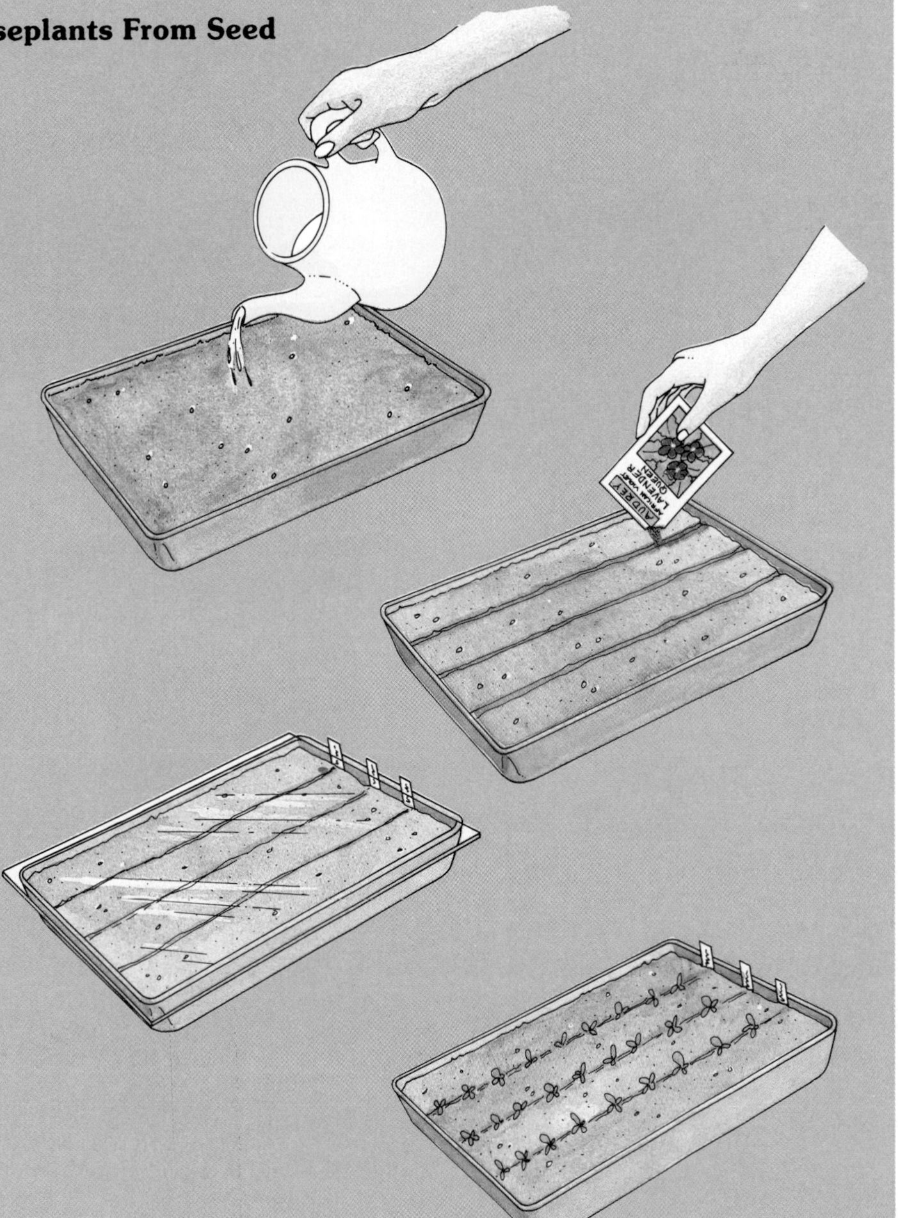

Pests and Problems

When you bring home new plants you should segregate them from all other plants for about a month, no matter where you got them. Nurseries and plant stores do not willingly sell sick plants, but problems can escape their notice. Isolating new plants keeps the problem from spreading. If you discover a sick plant in your collection, isolate it immediately and check nearby plants very carefully for signs of contamination.

If you need to use a pesticide, follow the directions precisely and always check the label to be sure the pesticide is suitable for the plant and the pest you are treating. Apply houseplant sprays formulated for indoor use.

CULTURAL PROBLEMS

Low Light

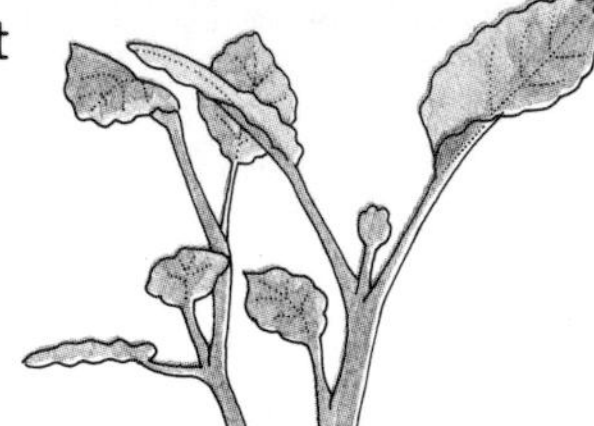

Problem: Plants will not flower. New growth is weak and spindly with large gaps between the leaves.

Solution: Gradually move the plant to a brighter location. Most plants will tolerate more light if kept well watered. If a brighter location is not available, provide supplemental electric lighting as described on pages 40–41.

Sunburn or Leaf Scorch

Problem: Dead brown patches develop on leaves exposed to direct sunlight, or leaf tissue may lighten or turn gray. In some cases the plant remains green, but growth is stunted. Damage is most severe when the plant is allowed to dry out.

Solution: Move plants that cannot tolerate direct sun to a shaded spot, or cut down the light intensity by closing the curtains against direct sun. Prune off badly damaged leaves, or trim damaged leaves and flowers. Keep plants properly watered.

Salt Damage

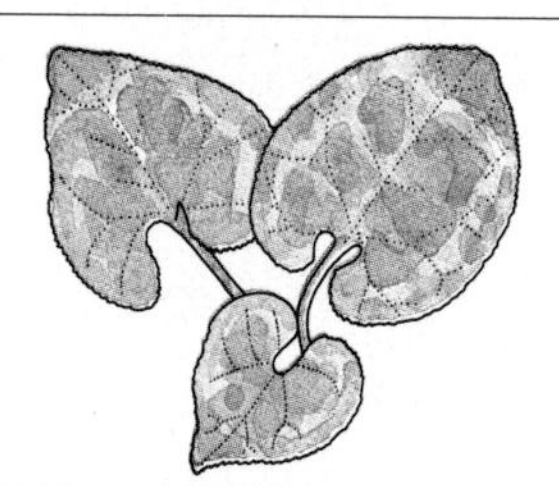

Problem: The leaf margins of plants with broad leaves or the leaf tips of plants with long, narrow leaves turn brown and die. This browning occurs on the older leaves first; but when the condition is severe, new leaves may also be affected. On some plants the older leaves may turn yellow and die.

Solution: Leach excess salts from the soil by flushing with water, as described on page 23.

DISEASES AND PESTS

Aphids

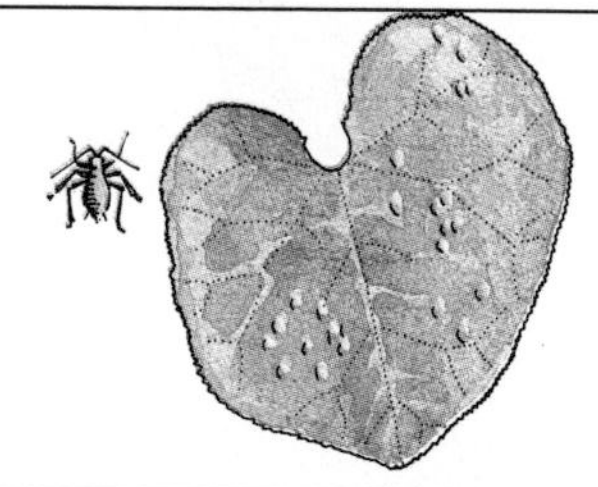

Problem: Leaves are discolored, curled, and reduced in size. A shiny or sticky substance may coat the leaves. Tiny, wingless, green soft-bodied insects cluster on buds, young stems, and leaves.

Solution: Wipe off small infestations with cotton swabs dipped in alcohol. A houseplant insect spray containing Orthene, resmethrin, or pyrethrins will control large infestations.

Botrytis Blight

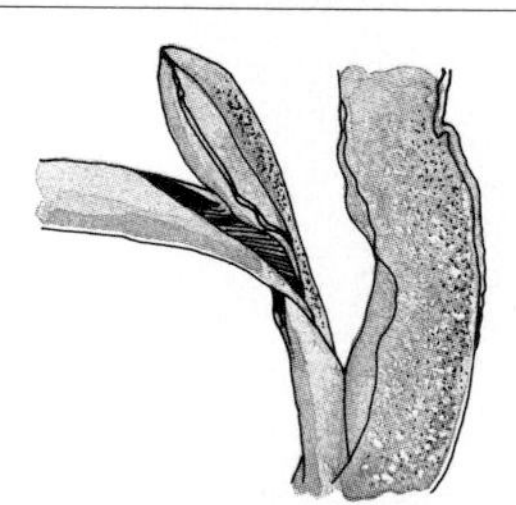

Problem: Brown spots and blotches appear on the leaves and possibly on the stems. Spots on flowers may be white, tan, brown, or purple, or the natural color of the flower may be intensified. If stems are infected, they may rot, causing the top of the plant to topple and die. Under humid conditions, the infected parts may be covered with fuzzy gray or brown growth.

Solution: Remove all diseased and dead plant material promptly, particularly old flowers. Avoid splashing water on the foliage and flowers, and avoid growing plants under crowded conditions where air is damp and still. Provide good air circulation around plants, but protect them from cold drafts. If the problem persists, grow plants in a warmer spot.

Crown and Root Rots

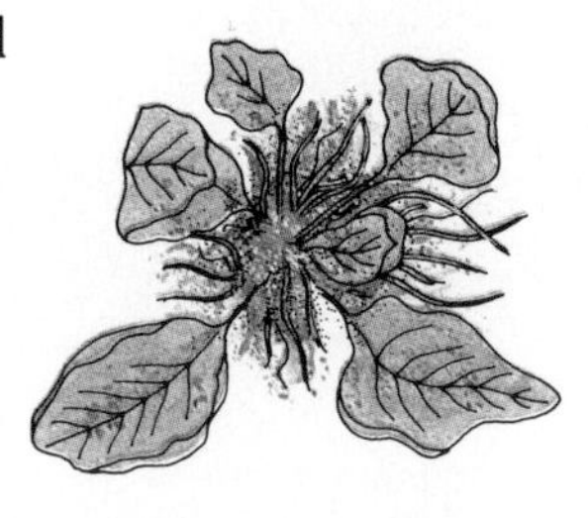

Problem: Leaves appear dull. Those in the center of the plant turn dark green and then black. This darkening rapidly progresses outward until the whole top of the plant is dead. The roots are dead and rotted.

Solution: There is no available chemical for eradicating this fungus once it gets established. Discard diseased plants, as well as the soil in which they grew. Soak pots for 30 minutes in a mixture of 1 part household bleach to 9 parts water. After handling infected plants, wash your hands thoroughly before touching healthy plants.

Cyclamen Mites

Problem: The newest growth in the plant center or stem tips becomes severely stunted. The leaves become brittle, stay very small, and may be cupped or curved. Color may change to bronze, gray, or tan. Flower buds fail to develop properly and open. Azaleas, begonias, cyclamen, kalanchoes, African violets, and many other plants can become infested.

Solution: Spray several times with a properly labeled dicofol miticide. Isolate mildly infested plants immediately. Discard badly infested plants. Soak the pots and wash the area where the pots were sitting with a solution of 1 part household bleach to 9 parts water.

Greenhouse Whiteflies

Problem: Tiny, winged insects feed mainly on the undersides of the leaves. The larvae are covered with white waxy powder. When you touch the plant, insects flutter around it. Leaves may be mottled and yellow.

Solution: If only a few leaves are infested, wipe off larvae with a damp cloth or cotton swab soaked in alcohol. Spray severe infestations with a houseplant insect spray containing Orthene, resmethrin, malathion, or diazinon. Check labels to determine which material can be used on your particular plant. Spray weekly as long as the problem persists. Remove heavily infested leaves as soon as the problem is spotted.

Mealybugs

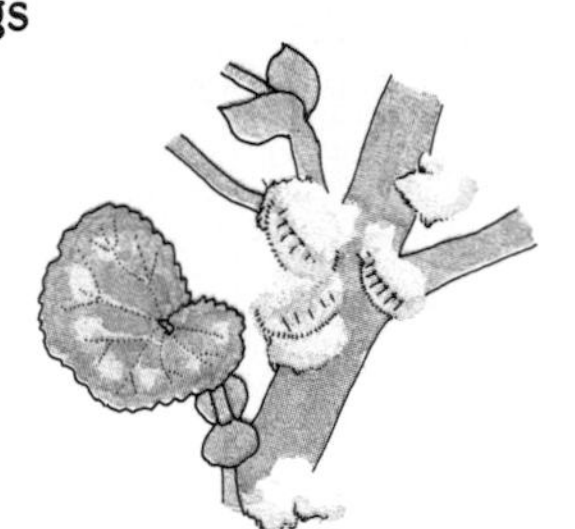

Problem: White cottony or waxy insects are on the undersides of the leaves, on the stems, and particularly in the crotches or where leaves are attached. The insects tend to congregate, resulting in a cottony appearance. Cottony masses that contain eggs of the insects may also be present. A sticky substance may cover the leaves or drop onto surfaces below the plant. Infested plants are unsightly, do not grow well, and may die if severely infested.

Solution: If only a few mealybugs are present, wipe them off with a damp cloth or cotton swab dipped in rubbing alcohol. Carefully check all parts of the plant to make sure all insects are removed. Wipe off any egg sacs under the rims or bottoms of pots. Discard severely infested plants and avoid taking cuttings from such plants.

Control with sprays is difficult because the waxy coverings on the insects and egg sacs tend to protect them. Spray stems and both sides of leaves with a houseplant insect spray containing Orthene or resmethrin and oil.

Powdery Mildew

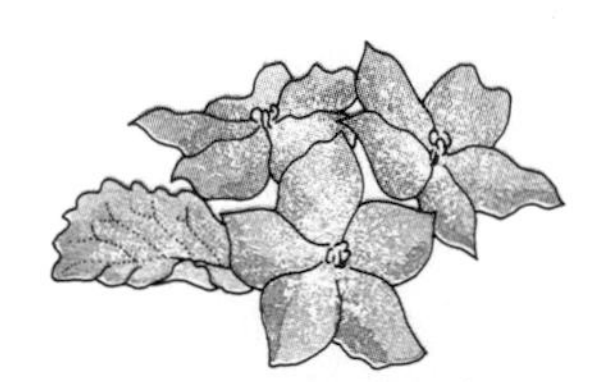

Problem: White or gray powdery patches appear on the leaves, stems, buds, and flowers. Leaves and flowers may be covered with the powdery growth. This material usually appears first on the upper surfaces of the older leaves. The affected plant parts may turn yellow or brown and shrivel up and die.

Solution: Spray with a fungicide containing dinocap or benomyl every two weeks until the disease is gone. Provide better air circulation around plant.

Scales

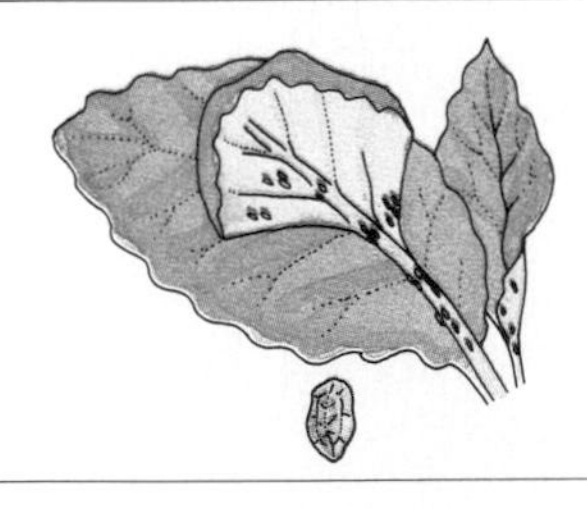

Problem: Nodes, stems, and leaves are covered with white, cottony, cushionlike masses or brown, crusty bumps or clusters of somewhat flattened reddish-gray or brown scaly bumps. The bumps are easy to scrape or pick. Leaves turn yellow and may drop. A shiny or sticky substance may cover the leaves.

Solution: Spraying is most effective against the crawlers rather than against the stationary adults. Spray with a houseplant insect spray containing Orthene or resmethrin and oil. Repeated applications may be necessary.

Spider Mites

Problem: Leaves are stippled, yellowing, and dirty. Leaves may dry out and drop. There may be webbing over the flower buds, between leaves, or on the lower surfaces of the leaves. To determine whether a plant is infested with mites, hold a sheet of white paper underneath an affected leaf and tap the leaf sharply. Minute green, red, or yellow specks the size of pepper grains will drop to the paper and begin to crawl around. The pests are easily seen against the white background.

Solution: Spray infested plants with an indoor plant insect spray containing Orthene, resmethrin and oil, or pyrethrins. Plants need to be sprayed weekly for several weeks to kill the mites as they hatch from the eggs.

Viruses

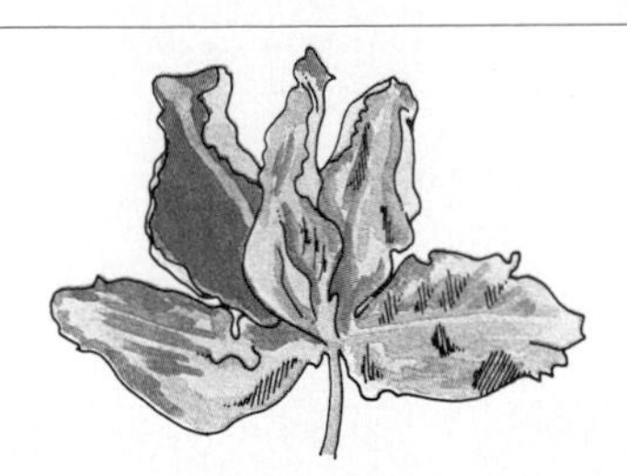

Problem: Foliage is streaked, mottled, or spotted with light green or white in an irregular pattern. The flowers may also be streaked or mottled.

Solution: There is no chemical control for viruses. To prevent the spread of the virus to healthy plants, remove and destroy infected plants.

Special window situations allow you to grow a wide variety of flowering houseplants. Find out how to maximize the light that is available in a window garden so that each plant receives the amount it needs to bloom properly. Learn how to arrange your plants so that the window garden is as decorative as it is functional.

The Window Garden

For every type of window, there is a type of garden that can be created in it. But an unplanned collection of plants in a window often makes a mess rather than a garden. However, it is possible to arrange plants near a window so that they not only grow well but also look their best.

In the early 1970s, window gardens were designed to obscure the window. The plants became draperies. The sill was full to overflowing and the glass was hidden by plants hanging from ingeniously knotted cords. Currently, real draperies are back in style and the window garden is a more sophisticated although less elaborate affair.

Rules of Window Gardening

The first rule for a genuine window garden is that all pots should match. They don't have to be exactly the same, but they should all be made of clay or all be the same color; they should harmonize. The aim is to draw attention away from the pots and focus it on the plants.

Rule number two is to think of the plants as a frame for the window, rather than a screen. Place tall plants at the edge of the window, or to one side. Medium-sized plants placed next to tall corner plants bring the greenery down to sill level. African violets mixed with begonias, cyclamen, kalanchoes, and gloxinias are all good for this intermediate level. Finally, something should "drip" below the sill level—a trailing African violet, or a small nematanthus. If you have a hanging plant, use it to replace one of the tall plants, rather than hanging it in the middle of the window.

WAYS TO USE WINDOWS

If the window is heavily draped on both sides, place one elegant flowering plant in an attractive pot or cachepot in the center and omit the collection.

If growing conditions on the windowsill are not ideal, use it as a presentation site and fill it from a collection of flowering plants kept in a plant-growing station elsewhere. If the plants are to be permanent residents, however, choose only the types of plants that will flourish in this window, eliminating those that, in your experience, will not grow well there. Remember to turn plants in a window occasionally to prevent them from developing a lopsided shape—caused when they lean toward the sun.

Bay windows offer more possibilities for plant designs. Build a pebble tray to fit the base of the bay and create a small conservatory. Use any combination of flowering plants that suit you and the window environment—flowering maples, lilies, begonias, orchids. Use seasonal plants during holidays. Stage plants at various levels in the window with shelves or plant stands. A rolled screen that separates the plants from the room is often used in such windows. If you are growing "short-day" plants, the screen can be lowered at night to provide the darkness they require to flower. If dark, cloudy weather is plaguing you and the screen is white or light colored, lower it to reflect light back into the garden and create more of a greenhouse effect.

Finally, such screens can protect plants from hot, dry blasts of air from the furnace during the winter.

Since flowering plants require more light than foliage plants, they usually suffer when grown far back from the window. However, damage or poor health may result from cold drafts or the burning rays of the sun if they are close to the glass. Filter summer sun with translucent curtains or choose a window that is naturally shaded by trees during the summer. Be aware that in the summer the high angle of the sun and the shade from trees can drastically reduce light that was strong in the

Pots for a window garden should be matched or, as here, concealed. Permanent foliage displays can be dressed up with seasonal bloom.

winter and spring. As fall approaches and the days shorten, the lower light intensity becomes especially noticeable before the trees lose their leaves. In the winter, although the lower angle of the sun provides more hours of direct sunlight in some windows, the sun is less intense and often obscured by clouds.

You can counter these variable lighting effects by moving plants as the seasons change. In addition, many indoor gardeners supplement window gardens with electrically produced light. It is a challenge to find a light fixture that will not ruin the appearance of your window. You can use valances or draperies to hide the lights. Track lighting with incandescent spotlights has also been used with some success.

The Skylight Garden

People who have skylights have a superb opportunity for a garden. Hanging plants do well under skylights because all the light is coming from overhead. Many flowering plants are well suited for skylights.

Hang plants at different levels around the sides of the skylight. If nothing is directly under the skylight, such as furniture or a passageway, you can hang a striking cascade of plants that reaches all the way to the floor. You might even want to build a large pebble tray to put under the skylight, creating a floor-level garden that mirrors the one above. A flexible arrangement for hanging plants can be made by attaching a metal grid or even a wagon wheel below the skylight.

Using only one type of flowering plant in a hanging garden creates a spectacular effect. Mistakes often occur when people try to combine plants. A striking display of six *Columnea* will draw more praise than a random crew of assorted plants. Try using several cultivars of the same plant, thus creating a bit of variation in flower color and form. Seasonally produced poinsettias in hanging baskets are becoming widely available. These baskets are produced with special growing techniques and are suited for temporary display only, as the poinsettia is not naturally a trailing plant. The trailing African violet is a relatively new form of African violet that is also becoming quite popular. The cultivars tend to be small, with flowers and leaves on long stalks. Several plants are usually planted in a single 6- or 8-inch basket.

Pots of trailing plants can be kept full, high, and bushy in the center by reintroducing small rooted cuttings of the parent plant into the main pot. Trailing African violets, however, have to be thinned out occasionally because the plants stop blooming if there is too much foliage. Hanging plants are best in plastic pots; hanging plants dry out quickly and the plastic pots retain water longer than the porous clay pots.

Tall dracaena contributes framing for this cheery window garden featuring bromeliads, heart-leaf philodendron, and asparagus fern.

Greenhouse windows create permanent presentation sites.

The Small, Attached Greenhouse

Prefabricated window greenhouses, small glass projections that replace standard windows, can be both a decorative space and a working greenhouse. Many of the same rules of regular window gardens apply—use compatible types of containers and plan out your plant design.

Small greenhouses that are attached to the home are also becoming widely available. These are constructed for many purposes: as an extra room, as a passive solar collector to lower heating costs, or simply as a way to grow more plants. Whatever the reason, a greenhouse will be more satisfying if it contains flowering plants.

You must plan properly if you intend to successfully manage a window or small attached greenhouse. Crowded, overgrown plants can quickly become a problem. They are an eyesore and they are also prone to health problems—pests and diseases spread rapidly in these circumstances. The greenhouse does not ensure that the plants will grow well and remain healthy. Diseases and infectious insects, nutritional imbalances, temperature and humidity deficiencies or excesses, and watering practices must be attended to properly.

Because the spaces are usually small, with less than a foot between shelves, window greenhouses are wonderful places to display a collection of small plants. By limiting the varieties to one plant family with similar characteristics, you automatically achieve a unified display. Depending on the exposure, you might try begonias or miniature geraniums (south-facing windows),

Light and temperature are controlled in this small attached greenhouse by opening a door to an adjacent lath house, as well as by opening panels in the lath house. Cacti are grown on the warmer and lighter upper shelves, orchids below.

African violets (north- or east-facing windows), or small bromeliads like *Tillandsia* species (east- or west-facing windows). Of course, not everyone wants to specialize, but this kind of collection is usually more effective than a hodgepodge of unrelated little plants scattered around the shelves.

To prevent harmful cold drafts, window or attached greenhouses should be made of insulated glass. If they are large, provide some sort of supplemental heat. Also, some means of drainage must be installed to dispose of excess water. Plants in window greenhouses can simply be placed in trays or saucers and moved to a sink when you leach the soil, but greenhouses must have some means of floor drainage. Fans or air turbulators may be needed to gently circulate the air. Ventilation is needed to control damp weather diseases and expel superheated air on hot, sunny days. Finally, some sort of shading must be available. Use rolled shade cloth or shutters if possible; otherwise, coat the glass with a whitewash compound especially designed for commercial greenhouses.

The Electrically Lighted Garden

In nature, African violets prefer the subdued light of the forest floor or rocky crevices. This makes them ideally suited to electrically lighted sites indoors. Such sites usually do not provide the amount of light needed by most flowering plants—unless the lights are used to supplement natural light. However, many African violet cultivars can be kept permanently under electric lights. Collectors of African violets routinely keep most of their plants in an artificially lighted garden.

There are two kinds of light gardens—working and decorative. A working light garden is a growing station where you can grow and bring plants into flower indoors. It can consist of a single 4-foot fluorescent tube on legs used for coaxing a few African violets into bloom, or it may be an elaborate multitiered cart capable of holding 50 or more potted plants. The working light garden functions like a greenhouse, where plants are prepared for temporary use in presentation sites around the house. Since the working light garden has its own source of energy, it does not have to consume window space. In fact, many are placed in basements or other underused spaces around the home.

The decorative light garden, on the

other hand, is a garden with presentation rather than growing as its primary objective. In the best of these, the lights are hidden with reflectors designed so that all the light is directed onto the plants and none spills out into the room. Because fluorescent lighting is not strong, a decorative light garden cannot be very deep and is best filled with small plants. A decorative light garden can be set into a bookcase, can brighten a hallway, or can sit on a kitchen counter with lights mounted under the cabinets. Keep the plants in individual pots so that you can easily tend them and move them around.

Many people attempt to set up a working light garden with a standard fluorescent light fixture, which has two fluorescent tubes side by side under a reflector. This concentrates the light intensity at the very center, while the edges and ends of the fixture receive less light. Plants grown at the edges or ends may not flower well and so must be turned regularly or rotated with those in the center.

In the long run it is far more practical to invest in light fixtures especially made for indoor gardening. The fluorescent tubes in such fixtures are slanted, with reflectors designed to provide more even, less center-oriented light. If you wish to build your own fixture for a working light garden, use three fluorescent tubes mounted 7 inches apart on a reflector board painted white. You will find that 4- or 8-foot fluorescent tubes are several times more efficient than shorter tubes.

A properly designed fluorescent fixture with two or three 40-watt tubes in a reflector will light a 2-by-4-foot growing area. Two such fixtures mounted parallel will illuminate a bench 3 by 4 feet. Two industrial fluorescent fixtures, 8 feet long and suspended side by side, will light a bench 3 by 8 feet. This intense light is especially necessary when growing plants that require strong light, such as the florist's gloxinia (*Sinningia speciosa*) and Cape primrose (*Streptocarpus* species). Many types of fluorescent tubes are available, including a number developed specifically for growing plants. An ideal combination is one warm white tube with one daylight or cool white tube. Grow-light tubes that emit a bluish purple light are also available. However, these may alter the apparent color of the foliage or flowers of your plants. Wide-spectrum plant lights can partially overcome this problem.

Whether you buy a light garden or build one yourself, be sure that it can be adjusted to stand or hang 12 to 24 inches above the surface on which the plants are placed. You should be able to move the light up and down to meet varied light intensity requirements. Plants in the garden that require stronger light can be placed on wooden blocks to move them closer to the light.

Fluorescent tubes should be lit 12 to 14 hours a day for African violets. Other plants may need up to 16 hours. Less time results in poor growth; more is harmful to plants. Most growers turn lights on at 7 or 8 a.m. and off at 10 or 11 p.m. Use an automatic timer for convenience and consistency.

Fluorescent tubes gradually decline in their light output with age. You may not notice this, but your plants will be affected. Replace fluorescent tubes once a year. Use a grease pencil or marking crayon to write the installation date on the end of each tube, so you know when to change them.

A working light garden allows dozens of plants to be grown in a small space.

THE LIGHTED TERRARIUM

With the advent of miniature African violets, florist's cyclamen (*Cyclamen persicum*), and other small flowering plants, the lighted terrarium has become popular as an indoor garden. Smaller-leaved flame violets (*Episcia* species) also do well in a terrarium, especially the pink-leaved 'Cleopatra'. If properly planned, the terrarium requires very little care and provides the warm, humid environment that many plants prefer. The light, however, is rarely as strong as that which would be provided by a standard electrically lighted indoor garden. As a result, flowering plants may have to be rotated in and out of a terrarium in order to keep them looking their very best. Keep flowering plants in their pots and bury them in the spots in the terrarium that you have reserved for blooming specimens.

Your terrarium may need some sunlight, but the sun also heats the plants. To prevent heat damage, do not seal a terrarium — lighted or not — that receives daily sunlight. An unsealed terrarium dries out more quickly, however. Keep a close eye on the moisture level so the plants do not become stressed.

A lit terrarium can house a landscape in miniature.

Flowering plants are a source of pleasure and beauty in a home, but the method in which they are displayed makes all the difference. This chapter teaches you how to display flowering houseplants so that they are shown to their best advantage in a setting that also allows you to care for their needs and prolong their flower display.

Displaying Flowering Plants Indoors

No matter how inherently beautiful a flowering plant may be, it must be properly displayed to be fully appreciated. There are a great many ways in which flowering plants can be used indoors. Interior designs and decorating styles vary widely, and so do the needs of the plants you display. There are, however, a few basic principles of proper plant presentation and display that you can learn and apply to your specific decorating situation.

The placement of flowering plants to enhance the decor of your home may or may not be coincident with the needs of those plants. Part of the green revolution in indoor environments that we spoke of earlier is due to a new understanding of the need to design and decorate with plants only in areas of the home that are optimum for plant growth. Nevertheless, indoor flowering plants are often placed in presentation sites selected according to the color, size, and form of a plant and its blossoms, with little regard to whether the plant could survive in the site. If you do this, you are treating the flowering plant in the same way you would an arrangement of cut flowers; the plants are a temporary decoration.

If you place flowering plants in spots where they cannot functionally exist for long, you must simply move them back to a growing station after you have enjoyed the flowers.

Start With a Good Plant

Skillful selection of a healthy plant that will bloom well is the first step in learning to decorate with flowering indoor plants. Purchase plants from reputable dealers and be wary of bargain plants. Develop an awareness of the way a healthy plant looks. Is the foliage the right size, the correct color, and evenly distributed throughout the branch structure? Is the plant well branched, free of stunting, and does it have a sufficient number of developing buds? It is the number of flower buds on a plant rather than the number of opened blooms that determines future performance of most flowering plants used indoors. Remember, start with plants that are beginning their flowering process.

Examine the foliage closely for signs of pests or diseases, especially powdery mildew, spider mites, and mealybugs. If possible, knock the plant from its container and examine the roots for pests or rot and check to see if roots are numerous and evenly distributed throughout the soil ball. Make sure the plant is wrapped properly for transit and do not allow it to remain wrapped or boxed for a long period of time. Unwrap it as soon as you get home.

If you are bringing your own plants into flower rather than buying plants at the onset of bloom, give your plants plenty of room to grow and prune or trim them into proper form before allowing them to begin the flowering process. As the flowering process approaches, withhold or cut back on fertilizer to plants other than African violets. Keep plants out of direct sun and drafts of hot air. Never let the soil dry excessively between waterings. Refrain from spraying with pesticides. Then follow the same general guidelines given above to ensure that you are starting with a healthy plant that is free of problems.

Left: This chrysanthemum would be a bad buy. It has been blooming in the shop until there are no new buds and the oldest blooms are beginning to brown and wilt. Look for plants just coming into bloom.

Right: Inspect plants in the store for pests and diseases. When you get them home, observe them closely for a few days, and keep them from touching your other plants during this time.

BRINGING PLANTS INDOORS

At the end of summer, many gardeners wish to bring frost-sensitive flowering plants indoors for an extended blooming season. Marigolds, begonias, impatiens, or petunias can be cleaned up, checked for pests, and planted in pots to decorate almost any room in your home. Generally, these plants have to be cut back to a form and size more compatible with your indoor setting. This is especially true with plants that are growing directly in the garden rather than in a container on the porch or patio. Garden plants that must be potted will suffer more shock than those already in a container. Pruning them back reduces this shock. Prune carefully and with discipline. An individual blossom may be particularly beautiful, but if it is on a spindly or drooping branch, remove it. If a few pests are found on the plants, apply the proper treatment a week or more before you move them indoors. Even then, do not place these plants near your permanent indoor specimens. One application of a pesticide may not kill all the pests present. The survivors will not cause any trouble as long as they stay on that plant. If the infestation is heavy, it is best to leave that particular plant outdoors and bring in a healthier specimen. If they are placed in a low-light indoor setting, regard them in the same manner you would cut flowers. Do not fertilize the plants you bring indoors and do not expect any growth or formation of new flower buds. If you place these plants in a greenhouse, you can expect continued growth and flowering for some weeks. Fertilization, continued pest control, and occasional pruning are important for outdoor plants moved into a greenhouse.

Using Color

The major decorative elements of a flowering houseplant are color, size, shape, and texture. Plants are often chosen to complement the colors and mood of an entire room, or of decorative items such as paintings, floor coverings, pillows, or sculpture. In such situations, it is important to plan for a succession of blooms throughout the year and it is often desirable to change the types of plants used. To do so expertly you need to become familiar with many types of plants.

The primary objective of decorating with flowering houseplants is to introduce new colors into the room. Color is a basic element of design because it is so easily perceived. Furthermore, colors carry familiar connotations of mood and spirit. Some colors are hot and exciting, others are brash and unsettling, and still others are cool and smooth. A brief review of some of the attributes of color will help you in your selection of indoor flowering plants.

GREEN

First of all, do not forget that one of the reasons for selecting a potted flowering plant is its foliage. Shades of green will always be a part of the color added to a room by a flowering houseplant. Greens generally promote a soothing, restful mood. Green will unify and pull together yellows and blues. Green isolates and calls attention to reds.

WHITE

White is the most reflective of all colors. Although it is a neutral color for decorating, the whites of blooming houseplants are commonly shades of cream, with accents of yellow, pink, or light blue. White blooms are most striking in large quantities, as in a blooming Easter lily. In groups of flowering houseplants, white improves contrast and creates highlights.

YELLOW

Yellow is a highly reflective color that can brighten almost any decoration. It is full of life and spirit and is especially welcome on indoor plants in the spring.

Green is the predominant color in all plants, even blooming ones, and can add as much to a color scheme as the flower color. Here impatiens repeats lamp base motif and greens of both lamp shade and carpet.

In flowers, white is often accented with a pastel color, as pink accents these white cymbidium orchids.

An orange Vriesea splendens *adds drama in an otherwise neutral room.*

Yellow can lighten the heaviness of large blue or purple objects. In fact, yellow will accent and complement any blue or green item. Yellows are also useful to warm earth tones, especially when reds would seem too hot.

ORANGE

Orange is a mixture of yellow and red. As such, it carries some of the feelings of both of these colors. Use it to combine the warmth of red and the liveliness of yellow. Orange brashly accents earth tones and can be used in small amounts to intensify or set apart the coolness of blues.

RED

Red is a strong, hot color. Red-flowered indoor plants do not have to be massive or have lots of flowers to be effective. The fact that small red Chinese hibiscus plants (*Hibiscus* species) generally produce only one bloom

Yellow flowers evoke bright spring sunshine in any season. Here shrimp plant (Justicia brandegeana) *and miniature primroses sing a harmonious duet.*

Above: Flamboyant red amaryllis is the star here, in a strong arrangement that works well because this kitchen has no competing motifs and/or colors. Use smaller splashes of red to accent a room where other patterns and colors vie for attention.

Right: Pink, though still a warm color, has a quieter effect than red. Pink azaleas set a warm, friendly tone in this otherwise spare room.

at a time is actually an attribute. Orange-reds are commonly used to blend or pick up highlight colors from fabrics or room accessories.

PINK

Pink is actually somewhere between the colors red and blue. Hot pink has the same effect as red, while bluish pinks are cool and formal in the same way that purple is. Pink is a softer color than red and can be used with reds to tone down their effect. Pink flowers convey a warm and friendly feeling.

PURPLE

Purple and lavender, like pink, fall between red and blue. Whereas pure reds are rather demanding, shades of

purple quickly become cooler and more formal. Lavender is softer than purple. Both colors tend to recede and are sometimes lost in a room, but they can be brought out if combined with white.

BLUE

Most plants that have blue flowers actually occur in shades of violet or blue-green. Thus some blues appear warm and some cool. Plants with blue blossoms can be blended with greens to create a spacious feeling. Try using flowering houseplants with blues where the use of a stronger colored flower might overwhelm the setting.

Above: Purples range from cool, deep blue purples to warm near-pinks. Most tones tend to recede in a room. Here lavender African violets stand out because they are grouped together and set off by white cachepots.

Left: Blue is generally a cool color, although blue-lavender is warmer than aqua. Lovely and restrained, Arabian violets (Exacum affine) *suit this formal setting without overpowering it.*

Some Indoor Flowering Plants Grouped by Color

WHITES

Agapanthus species — lily-of-the-Nile
Begonia × *cheimantha* — Christmas begonia
Begonia × *hiemalis* — Rieger begonia
Begonia × *semperflorens-cultorum* — wax begonia
Begonia × *tuberhybrida* — tuberous begonia
Camellia japonica — camellia
Carissa grandiflora — natal plum
Cattleya species — cattleya
Chrysanthemum × *morifolium* — mum
Citris limon — lemon
Clerodendrum species — glorybower (red bracts)
Crinum species — Bengal lily
Crocus species — crocus
Cyclamen persicum — florist's cyclamen
Eucharis grandiflora — Amazon lily
Euphorbia pulcherrima — poinsettia
Gardenia jasminoides — gardenia
Hibiscus species — hibiscus
Hippeastrum species — amaryllis
Hoya species — wax plant (pink center)
Hyacinthus orientalis — hyacinth
Impatiens wallerana — patient Lucy
Lilium longiflorum — Easter lily
Muscari species — grape hyacinth
Narcissus hybrids — daffodil
Ornithogalum species — star-of-Bethlehem
Pelargonium species — geranium
Primula species — primrose
Rhododendron species — azalea
Rosa species — miniature rose
Saintpaulia species — African violet
Senecio × *hybridus* — cineraria
Sinningia speciosa — florist's gloxinia
Spathiphyllum species — peace lily
Streptocarpus species — Cape primrose
Tulipa species — tulip
Zephyranthes species — zephyr lily

YELLOWS

Abutilon species — flowering maple
Achimenes species — rainbow flower
Aphelandra squarrosa — zebra plant
Begonia × *hiemalis* — Rieger begonia
Begonia × *tuberhybrida* — tuberous begonia
Calceolaria species — pocketbook flower
Capsicum species — ornamental pepper
Chrysanthemum morifolium — mum
Citrus limon — lemon (fruits)
Clivia miniata — kaffir lily
Crocus species — crocus
Episcia species — flame violet
Guzmania species — guzmania
Hibiscus species — hibiscus
Kalanchoe blossfeldiana — kalanchoe
Kohleria species — kohleria
Lachenalia species — Cape cowslip
Narcissus hybrids — daffodil
Nematanthus species — nematanthus
Primula species — primrose
Rosa species — miniature rose
Tulipa species — tulip
Zephranthes species — zephyr lily

ORANGES

Abutilon species — flowering maple
Anthurium species — flamingo flower
Begonia × *hiemalis* — Rieger begonia
Begonia × *tuberhybrida* — tuberous begonia
Capsicum species — ornamental pepper
Chrysanthemum morifolium — mum
Clivia miniata — kaffir lily
Columnea species — columnea
Crossandra infundibuliformis — firecracker flower
Episcia species — flame violet
Guzmania species — guzmania
Hibiscus species — hibiscus
Hippeastrum species — amaryllis
Impatiens wallerana — patient Lucy
Kalanchoe blossfeldiana — kalanchoe
Narcissus hybrids — daffodil
Nematanthus species — nematanthus
Pelargonium species — geranium
Rosa species — miniature rose
Schlumbergera species — Christmas cactus
Tulipa species — tulip
Zephyranthes species — zephyr lily

REDS

Abutilon species — flowering maple
Achimenes species — rainbow flower
Aeschynanthus species — lipstick plant
Anthurium species — flamingo flower
Begonia × *hiemalis* — Rieger begonia
Begonia × *semperflorens-cultorum* — wax begonia
Begonia × *tuberhybrida* — tuberous begonia
Calceolaria species — pocketbook flower
Camellia japonica — camellia
Capsicum species — ornamental pepper
Clerodendrum species — glorybower (with white)
Columnea species — columnea
Crinum species — Bengal lily
Crossandra infundibuliformis — firecracker flower
Cyclamen persicum — florist's cyclamen
Daphne odora — winter daphne
Episcia species — flame violet
Euphorbia pulcherrima — poinsettia
Guzmania species — guzmania
Haemanthus species — blood lily
Hibiscus species — hibiscus
Hippeastrum species — amaryllis
Hyacinthus orientalis — hyacinth
Impatiens wallerana — patient Lucy
Kalanchoe blossfeldiana — kalanchoe
Kohleria species — kohleria
Lachenalia species — Cape cowslip
Pelargonium species — geranium
Primula species — primrose
Rhododendron species — azalea
Rosa species — miniature rose
Schlumbergera species — Christmas cactus
Senecio × *hybridus* — cineraria
Sinningia hybrids — gloxinera
Sinningia speciosa — florist's gloxinia
Solanum pseudocapsicum — Jerusalem cherry (fruits)
Sprekelia formosissima — Aztec lily
Tulipa species — tulip

PINKS

Abutilon species — flowering maple
Amarcrinum memoria-corsii — crinodonna
Begonia × *cheimantha* — Christmas begonia
Begonia × *semperflorens-cultorum* — wax begonia
Begonia × *tuberhybrida* — tuberous begonia
Billbergia species — vase plant
Calceolaria species — pocketbook flower
Camellia japonica — camellia
Crinum species — Bengal lily
Cyclamen persicum — florist's cyclamen
Daphne odora — winter daphne
Episcia species — flame violet
Euphorbia pulcherrima — poinsettia
Haemanthus species — blood lily
Hibiscus species — hibiscus
Hippeastrum species — amaryllis
Hyacinthus orientalis — hyacinth
Hydrangea macrophylla — big-leaf hydrangea
Impatiens wallerana — Patient Lucy
Kalanchoe blossfeldiana — kalanchoe
Pelargonium species — geranium
Primula species — primrose
Rhododendron species — azalea
Rosa species — miniature rose
Saintpaulia species — African violet
Senecio × *hybridus* — cineraria
Sinningia speciosa — florist's gloxinia
Streptocarpus species — Cape primrose
Tulipa species — tulip
Zephranthes species — zephyr lily

PURPLES OR LAVENDERS

Crocus species — crocus
Primula species — primrose
Saintpaulia species — African violet
Senecio × *hybridus* — cineraria
Sinningia hybrids — gloxinera
Streptocarpus species — Cape primrose
Tillandsia species — blue torch
Tulbaghia species — society garlic
Veltheimia species — forest lily
Vriesea species — vriesea

BLUES

Achimenes species — rainbow flower
Aechmea species — living vase plant (large pink bracts)
Agapanthus species — lily-of-the-Nile
Crocus species — crocus
Exacum affine — Arabian violet
Hyacinthus orientalis — hyacinth
Hydrangea macrophylla — hydrangea
Muscari species — grape hyacinth
Primula species — primrose
Saintpaulia species — African violet
Scilla species — squill
Senecio × *hybridus* — cineraria
Streptocarpus species — Cape primrose
Tillandsia species — blue torch

Finding a Place for Your Plant

Some flowering plants can be intermixed with other indoor plants to add color to an indoor garden. Other flowering plants are so remarkably beautiful that they should be placed by themselves in a presentation site such as a dining table, entranceway chest, or a side table in front of a mirror. If the station is in a formal area of the house, the plants themselves must be reasonably symmetrical. If more than one plant is used, their arrangement must be symmetrical as well.

You do not have to restrict your display of indoor flowering plants to prominent plant stations. You can use flowering plants indoors in almost as many situations as foliage plants. The only difference is that massive indoor garden settings are usually designed with a combination of foliage and flowering plants. Flowers are color, and color provides accent, especially when used against a backdrop of the greens of foliage.

PRESENTING YOUR PLANT

As you read earlier, the way that flowering plants are presented can make the difference between "decor" and mere clutter. Decorating with African violets, for instance, is no easy job. The plants are small, and avid collectors often have hundreds of them covering every available surface or crowded into fluorescent light gardens. Individually, African violets are quite beautiful, but they need proper presentation in order to be decorative.

Take a single small plant and slip its plastic pot into a small cachepot. Set the cachepot on a round stand that just fits the bottom of the pot. Instantly the African violet turns into a star.

Two important principles of display are illustrated here. First, the "mechanics" must not show. The green plastic pot with the rolled edge that African violet growers prefer is one part of the mechanics. It serves its purpose, but is not lovely to look at. The decorative cachepot is part of the presentation (and since it has no hole in the bottom, it doesn't drip water on the table). Second, whatever you display must look important. If you isolate a single plant from its fellows, spotlight it, put it in a showy pot, and set it on a stand, it instantly becomes more important and eye-catching.

In this example, the stand and the cachepot combine to create a plant station that can be placed wherever it looks best. Blooming African violets can be rotated from their growing areas into the presentation site. Flowering plants that are dressed up and displayed one at a time provide far more enjoyment than plants that are left perpetually in group arrangements, some flowering, some not, all in plastic pots and all too often set atop water reservoirs made from margarine containers — the antithesis of an aesthetic display.

Of course, African violets can look very attractive when they are presented in small groups. Plant stands that hold half a dozen or so small plants are often used as presentation sites for African violets. The miniature or standard-sized trailing varieties are attractive when combined into a hanging planter. When grouping together individual containers of flowering plants, plan to vary the shades of flower color, and arrange them so the flowers are on different levels for eye appeal. The art of placing plants at varying elevations is called staging. It is a common practice in flower shows.

Many indoor flowering plants have sensational blooms. Thus they are commonly used in spite of the fact that they may have a rather gross form or drab foliage. Some orchids and bulbous plants fall into this category. Even though a plant's blossom is so exquisite that you feel its negative points can be pardoned, others will notice the rank foliage or straggly form. Be careful with the presentation of these plants. Hide the unsightly portion of the plant by grouping other plants around it.

Left: Here the problem of unattractive leaves is solved by camouflage. A large spathiphyllum has been placed so that it appears to be foliage for the yellow orchid to the left of the window. Above: The jewel-like beauty of a royal purple African violet is accentuated by a handsome cachepot and plant stand.

Holiday Displays

Exceptions to many of the above concepts abound when plants are part of a seasonal or holiday display. Masses of blooming plants in unorthodox presentation sites can add holiday excitement to your home. Poinsettias placed on a staircase at Christmas or chrysanthemums massed on a hearth at Thanksgiving are perfectly acceptable whatever the color scheme or practicality. Use strong, primary colors or whites, and mass the plants, do not crowd them. It may be best to remove permanent plant residents for the temporary holiday display. Finally, do not prolong the display long after the holiday has passed. Be careful to buy plants so that their flowering peaks on the holiday. If you misjudged the timing, you might be happier if you enjoy the plants you have in some other part of the house and buy another set of plants for your intended display.

Prolonging the Flowering Period

Your primary object in caring for your plants while they are flowering is to prolong the flowering process for as long as possible. It is not uncommon for indoor gardeners to double the usual life of blooming plants in a decorative indoor setting. Most plants that have been properly grown prior to their flowering period will have enough nutrients to carry them through the period when they are being displayed. Additional fertilization, especially without thorough watering, may actually damage the roots of such plants. Never allow the rootball to dry excessively. The new organic growing media used by many commercial growers is especially difficult to rewet once it dries out. Excessive drying and root damage causes sudden leaf or bud drop in many plants. Finally, keep most blooming plants out of direct sunlight or drafts of hot, dry air. The flower petals cannot replenish lost water as easily as leaves. Petals may burn, fade, or wilt under hot, dry conditions.

Poinsettias say Christmas; use them to flank a fireplace or line a stair. To buy long-lasting poinsettias, check in the center of the brightly colored bracts for the small flowers. If these are still in bud, the plants will stay bright longer.

PLANT CARE AFTER FLOWERING

Unless you plan to discard the plant toward the end of its flowering period, you must change your care routine when the plant stops flowering. This is most easily accomplished by returning the plant to its permanent growing station. In the case of plants that require a rest immediately after flowering, begin withholding water. Do so gradually over a period of a few weeks and do not fertilize during this time. Withhold water by watering less frequently, not by using less water at each irrigation. Many bulbous plants need their foliage to continue to produce food, which is stored in the bulb for the next flowering cycle. Yellowing foliage naturally signals the onset of their dormancy.

For successful decorating with flowering plants in presentation sites, it is important to discard or replace a plant that is past its prime. Commercial interior landscapers call indoor flowering plants "changeout plants" because they realize that they must keep rotating plants that flower if the site is to be constantly decorated by flower color. It is easy to tell when a plant is past its flowering prime. These plants will not "come back" into bloom unless they are pruned and possibly repotted. Furthermore, the plants generally have to be moved to a growing station to bring them back into bloom. When the flowers are fading and no new buds are forming, move the plant to a growing station and put a new specimen that has just begun to flower into the presentation site.

Chrysanthemums are among the plants best used as seasonal dressing. Here a third basket is reserved to add seasonal bloom in a foliage plant arrangement.

Renda Hildebrand

Over 80 different types of flowering houseplants are described in this chapter, beginning with an in-depth discussion of the most famous and ever-popular flowering houseplant, the African violet. All the information you need to bring each of your houseplants into bloom and keep them blooming for as long as possible is included.

An Encyclopedia of Indoor Flowering Plants

In this chapter you will be introduced to more than 80 widely available flowering plants suitable for growing or using indoors. Review the entire section to become familiar with its general organization and layout. You'll find it gives specific cultural facts and environmental requirements, descriptions of growth and flowering habit, and photographs to help you select plants that best fit your needs and personal preferences.

With the exception of African violets, the encyclopedia is arranged in plant groupings based on a combination of growth and blooming habits. African violets are placed at the beginning of the chapter because of their star status among flowering houseplants. These are the plant groupings and the order in which they appear in the encyclopedia: African Violets (this page); Plants That Bloom Continuously Indoors (page 64); Plants That Bloom Seasonally or Intermittently Indoors (page 68); Plants That Need a Rest After Blooming (page 74); and Plants Used Indoors Only While in Flower (page 85). If you have difficulty locating a particular plant, use the index at the back of the book.

The description and care guides are written with an emphasis on care for the plant and its appearance when you are inducing it to flower or when it is in flower. The light requirements are those that are best for the plant while it is in flower. If the plant needs a different light level to bring it into flower, this is also noted. In some cases, light requirements for the plant after it flowers are included because these are particularly important in order to bring about another flowering cycle. Fertilizer requirements are given for continued culture of the plant. As noted earlier, many flowering plants need less fertilizer during flowering. The other care guides are presented with the assumption that you plan to keep the plant in your collection after it has flowered.

SAINTPAULIA SPECIES—AFRICAN VIOLETS

African violets are not related to the familiar woodland violets. They are members of the large Gesneriad family, which includes the florist's gloxinia (*Sinningia speciosia*) and the lipstick plants (*Aeschynanthus* species). Today, African violets are the most popular and widely grown gesneriads of all, and the most popular flowering houseplant of all time.

African violets were first discovered by European colonists in Africa less than a hundred years ago. Seeds were sent to Ulrich von Saint Paul-Illaire in Germany by his son, then governor of an East African German colony (now part of Tanzania). The Baron gave the seeds to the prominent botanist Herman Wendland, the director of the Imperial Botanic Gardens at Hanover. In two years Wendland flowered these new plants and had the thrill of describing not only a new species, but a new genus as well. He named the plant *Saintpaulia ionantha. Saintpaulia* honors the family of the Baron, while *ionantha* is Greek for "violet-like" or "violet-colored."

In 1893 these new plants were shown by Wendland at the International Horticultural Exhibit in Germany. Wendland's *Saintpaulia* species created great excitement. Within a few years of their debut in Germany the African violets, as they quickly became known in common usage, sparked interest in the horticultural community all over Europe. The possibility of having flowers in the house during the winter (and possibly all year) was quite a selling point. Unfortunately, after this first wave of popularity, enthusiasm waned and African violets became practically unknown for several decades.

A New York florist introduced African violets to this country in 1894, but their initial reception was rather cool—both literally and figuratively. More often than not, the plants took a chill and rotted or died in the drafty homes of late-eighteenth-century America, and African violets quickly developed a reputation as difficult and finicky.

Until this time no conscious hybridization program had been developed. In 1926 Armacost and Royston Greenhouses of Los Angeles, today known for their orchid work, decided to take a chance on African violets. They imported seeds from Europe (seeds of what was then named *S. ionantha* × *S. confusa*), grew the plants, and, when they flowered, selected what they considered to be the 10 best. These 10 were named and propagated vegetatively in large numbers.

By 1936 a crop was ready for sale. The 10 plants were 'Blue Boy', 'Sailor Boy', 'Admiral', 'Amethyst', 'Norseman', 'Neptune', 'Viking', 'Commodore', 'No. 32', and 'Mermaid'. Though all were single-flowered and in a narrow color range, they were good plants. They flowered well and had handsome foliage. Many of these cultivars are still grown by hobbyists today.

After just a few years Armacost and Royston gave up propagating African violets, but fortunately a few botanical gardens and commercial growers kept growing them. One was Roger Peterson, son of a Danish immigrant and greenhouse owner in Cincinnati, Ohio, who saw the plants in Fairmount Park Greenhouses in Philadelphia. In 1910 he purchased 100 plants and brought them to Ohio to continue their production and hybridization.

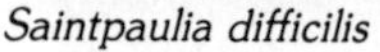
Saintpaulia difficilis

Saintpaulia ionantha

Mutations appeared in cultivation, widening the color range and creating new types of foliage. Flowers were larger, many petaled, and sometimes bicolored. Foliage became variegated, ruffled, or large and lush. The plants slowly gave up their cultural secrets, and seedlings that displayed exceptional vigor and ease of flowering were used in further breeding. Suddenly, they had become easy to grow.

In the 1930s, growers realized that African violets would adapt well to electric light gardening. Fluorescent lights became available at this time and may have been responsible for the renewed popularity of the plant. The truly unique adaptability of African violets to almost continuous bloom under such conditions, along with the growing variety in foliage and flower form and color, made them irresistible to indoor plant enthusiasts.

In 1946 interest had grown so much that the first national African violet show was held in Atlanta, Georgia. The attendance was so unexpectedly high that the police were called in to control the crowds and traffic! The African Violet Society was founded at this time. Interest has continued to grow steadily up to the present, and affiliated clubs of the African Violet Society can be found in any major city and many towns across the country.

Most of our modern plants are hybrids or plants that resulted from crossing two species. In most cases, the parent plants were *Saintpaulia ionantha* and *Saintpaulia confusa.* The best of their offspring were then crossed with each other or with their parents. In this way breeders developed the plants we see today.

Modern African violets come in a wide range of colors and leaf types. Early African violets came in shades of purple and blue, but they now come in almost every color except yellow and a clear red. Although hybridizers are working on red, the current reds still contain some blue.

The form of the plants has been improved from earlier cultivars: The flowers of earlier plants were hidden in the foliage or drooped on weak stems. But in the 1950s, patented lines and cultivars were developed in Germany; and they were released in this country in the 1970s. They all have full clusters of blooms held above the center of the plant on strong stems. Widely available today, the four most popular lines are Rhapsodie, Optimara, Diana, and Ballet. Each of these lines has many cultivars within it. (Remember, patented cultivars

Saintpaulia grotei

Saintpaulia orbicularis

An assortment of miniature African violets. Left to right: (back row) 'Irish Angel', 'Snuggles', 'Cloud 10', 'Skagit Toy Asset'; (front row) 'Baby Blue', 'Fickle Fairy', 'Hug a Lug', 'Bohemian Sunset'.

may not be propagated except by permission of the patent holder.)

The simple, rounded leaves of the early plants have also undergone a metamorphosis at the hands of the hybridizers. Cupped, fluted, and convoluted shapes with different patterns of variegation have been created.

Many cultivars bred by amateurs are registered with the African Violet Society, which keeps a Master List of Cultivars. Each is assigned a number and is described. A supplement to this list is published each year in *African Violet* magazine. Many descriptions of African violet cultivars in this book are based on this Master Variety List.

Here are a few of the most popular species being grown by hobbyists:

SAINTPAULIA CONFUSA: Clusters of single blue-violet to violet flowers sit atop medium to light green foliage. The leaves are sparsely to densely hairy, and slightly quilted. This species tends to grow at an angle and branch to produce several crowns (a crown is an individual rosette).

SAINTPAULIA DIFFICILIS: Flowers are shades of blue above long pointed leaves. The leaves often become "spooned" (curved up at the edges). The plant has long upright leaf stems and grows as a single crown. The veins of the leaf are prominent.

SAINTPAULIA GROTEI: This is one of the trailers, with stems that can reach over 3 feet in length. The leaves are flat, medium green, and have distinctive brown leaf stalks. The flowers are light blue with a dark center. This plant needs excellent drainage and will not tolerate overwatering.

Saintpaulia tongwensis

Saintpaulia 'Snowy Trail' (a semi-miniature trailer)

Saintpaulia 'Amigo' (boy foliage)

Saintpaulia 'Lynn' (Melody series) (girl foliage)

SAINTPAULIA INTERMEDIA: This plant is intermediate between upright and trailing forms. Sometimes it will grow upright as a single or multiple crown. The leaves are olive green, almost round, and have a small-toothed edge. The plant flowers freely, producing bunches of blue blossoms.

SAINTPAULIA IONANTHA: This species led to the widespread popularity of African violets in the United States. It generally has blue-violet flowers, though other forms exist. The leaves are dark green, quilted, and toothed on the edge. Usually this plant grows from an upright single crown.

SAINTPAULIA ORBICULARIS: The leaves on this species are almost orbicular, hence the name. The flowers are numerous and pale in color.

SAINTPAULIA PUSILLA: One of the species used to breed the miniature hybrids. It was a true miniature, the mature plant being only 5 inches across. Unfortunately, it has not been seen since before World War II, and is probably extinct.

SAINTPAULIA SHUMENSIS: This delicate-looking, semialpine plant has smaller, paler blue flowers than any of the other species. This is the species from which the recent miniature cultivars are being developed.

SAINTPAULIA TONGWENSIS: This species is very similar to *Saintpaulia ionantha*. It is free-flowering with single, lavender flowers. It must have high humidity to grow well.

Miniatures

The miniature and semi-miniature African violets are rapidly gaining in popularity. Many fine cultivars are now available with as much variation in leaf and flower form as the standards. Some miniatures have flowers almost as large as some standards. Forms even exist with bell-shaped flowers. Since the miniatures stay under 6 inches in width, and the semi-miniatures under 8 inches, a large collection can be housed in a small space. Their culture is essentially the same as their larger cousins, though some growers feel they need stronger light.

Trailers

The trailing African violets have been bred from those species that produce long stems, such as *Saintpaulia grotei*. This line of breeding is still in its infancy, but it has been improved greatly over the last few years. The trailing miniature African violets have small leaves and flowers. Grow them in a

Saintpaulia 'Pink Lemonade' (variegated foliage)

Saintpaulia 'Blue Excitement' (red reverse foliage)

pot or hanging basket no larger than 4 inches in diameter. Plant one or two trailing African violets in a pot. If the plants do not begin to branch out on their own fairly early, pinch out the center to hasten the process. Trailing African violets grow much faster than their regular cousins, and with proper light they make a stunning show with an effect that is entirely different from the single-crowned, wagon-wheel-shaped African violet. The mature trailing plant actually resembles a globe, totally covering the container. The many crowns may display hundreds of blooms on a continuous basis. Thin mature trailing plants occasionally by removing older large leaves. These older leaves tend to cover up the new crowns, which need light to bloom. Remove a few crowns from time to time to keep the plant blooming steadily.

LIGHT: Place in bright, indirect light from a south-, east-, or west-facing window. In winter, supplement daylight with artificial light.

WATER: Keep evenly moist. Water thoroughly and empty drainage from saucer. Keep water off the foliage, but wash plants with tepid water a few times a year. After washing, keep the plant out of direct sunlight until it is dry again.

HUMIDITY: Average indoor humidity.

TEMPERATURES: 55° F to 60° F nights, 70° F to 75° F days. Keep out of cold drafts.

PROBLEMS: Mushy, brown blooms and buds indicate botrytis blight. Pick off diseased parts. Provide good air circulation and avoid high humidity.

Yellow rings on leaf surface are caused by cold water touching foliage. Use tepid water only; cold water will cause leaf-spotting.

Streaked, misshapen leaves with irregular yellow spots are caused by a virus; there is no effective cure—discard plants.

If a healthy plant wilts suddenly, it may be developing the fungus known as crown rot, which results from erratic watering or overwatering. Allow the soil to dry out slightly, but not completely, and then soak it. After this, maintain a constant level of soil moisture. Repot plants that recover and discard rotted plants and the soil in which they were growing. Soak all pots that contained infected plants in a solution of 10 percent household bleach for 30 minutes prior to reuse. Severe temperature changes may also favor crown rot.

Lack of flowers is probably caused by inadequate light. Supplement daylight with artificial light. Very dry air or very cold air are also possible causes. Repotting and moving the plant to a new location can

Saintpaulia 'Mister Brad' (quilted foliage)

Saintpaulia 'Hot Toddy' (serrated foliage)

inhibit flowering for a long time.

Yellowing leaves result from dry air, too much sun, or incorrect watering. Improper fertilizing can also yellow the leaves, so follow directions carefully.

Brown, brittle leaves develop from soil that is deficient in nutrients or high in soluble salts. Repot if soil is old; otherwise, fertilize regularly and always be sure water drains from the pot when you water.

Slow growth and downward-curling leaves may be caused by low temperatures.

Brown-edged leaves and small flowers are a result of low humidity. Place plants on humidifying trays.

Further information on problems common to African violets is on pages 34–35.

PROPAGATION: Take leaf cuttings at any time. Suckers (side shoots) should be removed from the plants, or they will become unsightly and bushy. These suckers are easily rooted to make a new plant. A plant with a multiple crown when you purchase it should be divided as soon as possible.

GROOMING: Pick off yellowed leaves. For constant bloom, prune outside leaves if there are more than five rows on the plant.

REPOTTING: Keep plants slightly pot-bound. African violets can be repotted at any time, but it is best to avoid doing so in hot weather. Fall or spring are optimum for repotting.

FERTILIZATION: Fertilize all year, preferably with a constant feed program using a fertilizer designed for this purpose.

Saintpaulia 'Buttercup' (fringed foliage)

Saintpaulia 'Sarita' (strawberry foliage)

Saintpaulia 'Tsunami' (a microminiature)

Saintpaulia 'Vermont' (Optimum series) (plain, single flower)

Saintpaulia 'Sabrina' (Rhapsodie series) (plain, single flower)

Saintpaulia 'Wineberry' (semi-double flower)

Saintpaulia 'Starshine' (regular star flower)

AFRICAN VIOLETS: FOLIAGE AND FLOWER TYPES

Foliage Classifications

Plain, sometimes called "Boy" foliage—typical simple leaf, named for 'Blue Boy', the most famous of the earlier varieties.

Example cultivars:

'Heaven Sent'—double pink, fine white edge

'Susie Sport'—double purple

"Girl" foliage—named for 'Blue Girl', features wavy edges and a white spot at the base of the foliage. These are much rarer than true plain foliage types, as most girl foliage plants have a tendency to bunch up, especially if grown under improper light. A very attractive plant when well grown.

Example cultivars:

'White Madonna'—(one of the oldest and most widely known girl foliage plants) double white star

'Lady of Spain'—dark red double

Optimara cultivars (named after cities)

Variegated green foliage—becoming exceptionally popular. The newer cultivars bloom just as profusely as their green-leaved cousins and are available in the same wide range of blossom colors.

Example cultivars:

'Apricot Frost'—double apricot-pink bloom

'Frosted Ruby'—dark red star with white edge

Leaves with red reverse—reddening is particularly visible from the underside, very dark on top, some nearly black. Makes striking contrast against bright-colored or white flowers.

Example cultivars:

'Cameo Queen'—large double white

'Moonlight Maiden'—bright double pink stars

Oak leaves, with indented margins—tend to grow very large, and are therefore not as easy to handle for the average grower. People interested in growing large show plants may collect these.

Quilted leaves—markedly raised areas between veins. Tend to grow very large and are therefore not as easy to handle for the average grower. People interested in growing large show plants may collect these.

Example cultivars:

'Pink Cargo'—deep pink double

'Snow Prince'—white single

Serrated leaves—very noticeable teeth on the leaf edge.

Example cultivars:

'Jason'—deep raspberry pink, blue overcast

'May Dance'—pink and white semi-double to double stars

Fringed leaves—extremely rippled and serrated, giving a lacy look to the plant. Leaves tend to be twisted and turned. As these plants are almost impossible to grow into a perfect wagon wheel, they are no longer widely grown.

Example cultivar:

'Angel Lace'—white with blue markings

Strawberry foliage, sometimes called Holly leaf violets—have a waxy or shiny appearance.

Example cultivars:

'Harvey'—(the oldest known of this type)

'Ladies Touch'—pink and white

'Lilac Morn'—dark lilac

Microminiature—foliage may be green or variegated, but the entire crown will be up to 3 inches across with blossoms approximately ¼ inch in size.

Example cultivars:

'Pip Squeek'—tiny, pink bell-shaped flowers

'Tidee Bug'—reddish lavender, double flower

Miniature or semi-miniature—foliage may be green or variegated. Fully grown miniatures will be under 6 inches in diameter, semi-miniatures will grow to 8 inches across.

Example cultivars—Miniatures:

'Snuggles'—bright pink double

'Midget Valentine'—single fuchsia red

Example cultivars—Semi-miniatures:

'Baby Blue'—light blue double

'Irish Angel'—double light blue with a green edge

Trailing African violets—foliage is quite flexible. One plant produces many crowns, which eventually cover the top and sides of the pot.

Example cultivars—Miniature trailers:

'Pixie Blue'—single purplish blue with a slightly deeper center

'Happy Trails'—fuchsia pink double star

Flower Types

Plain single (pansy-shaped) blossom—most new single-blossom, plain-flowered hybrids are popular because they do not drop their flowers the way the older cultivars did. Many have an almost invisible extra petal near the center. Seventy-five percent of the new single-blossom hybrids are star-shaped.

Saintpaulia 'French Lilac' (double-fringed star flower)

Saintpaulia 'Pixie Blue' (a trailer)

Saintpaulia 'Fantasy Sparkle' (fantasy-type flower)

Example cultivars:
'Blue Power' — medium blue
Optimara cultivars — many colors

Semi-double blossom — some petaloids, stamens are still clearly visible.

Example cultivars:
'Camelot Rose' — dark orchid semi-double with a darker edge
'Remember When' — semi-double two-tone lavender

Double blossom — medium- to extra-full-sized petals. Double pansy-type flowers are rarely grown anymore.

Regular star blossom — flowers are a true five-pointed star. Occasionally forms a six-pointed flower.

Example cultivars:
'Dune Flower' — single fuchsia star
'Wonderland' — light blue

Double star blossom — medium- to full-petaled flowers, by far the most popular of all doubles.

Example cultivars:
'Surprise' — double white and blue star
'Star Wars' — double white star, wide violet edging

Double fringed blossom — double flowers with fringed petals. Generally star types.

Example cultivars:
'French Lilac' — lavender and wine
'Charlie' — double fringed purple

Bicolor — same basic color in the bloom, but slightly to heavily shaded.

Example cultivars:
'Janet' — pink and reddish
'My Stars' — double pink star with lavender band and overlay

Fantasy or multicolored blossom — flecks and dashes or streaks in colors different from the basic color. Can be single- or double-flowered.

Saintpaulia 'Kiss T' (bicolor flower)

Example cultivars:
'Fantasy In White' — white semi-double with blue and pink splashes.
'Picasso' — soft lavender double, spotted and streaked with purple.

Geneva-edged blossom — white edge all around; petals may be any type of flower.

Example cultivars:
'Like Wow' — royal purple star, white edge
'Captain Flash' — single pink, wavy white edge

Other edged blossoms — also called Picotee, white or light shades, with contrasting dark edge. Some have slightly ruffled edge.

Example cultivars:
'Roly Poly' — frilly white double, red wavy edge
'Circus Ring' — white semi-double, purple edge

Striped blossom, sometimes called Chimeras or Pinwheels — each of the semi-double petals has a distinct shading of a similar or contrasting color in the center of the flower petal. These cannot be cultivated by means of leaf cuttings. They must be reproduced from offshoots (suckers).

Example cultivars:
'Valencia' — bluish colors
'Mauna Loa' — deep red orchid center stripe, light rose border

Wasp blossom — single or double flowers. The shape resembles a wasp. Truly a collectors item, but still hybridized by some.

Saintpaulia 'Canadian Sunset' (fantasy-type flower)

Saintpaulia 'Tidecrest' (other edged flower)

Saintpaulia 'Spinner' (Geneva-edged flower)

PLANTS THAT BLOOM CONTINUOUSLY INDOORS

Although African violets are the best known of the indoor flowering plants that bloom almost continuously indoors, several other plants also have this remarkable capability. Many of these plants are gesneriads, but others range from seeded herbaceous plants to woody vines or small shrubs.

While no plant will literally bloom "continuously," the plants listed on the following pages can come close. If they are propagated and planted correctly, if they are grown into maturity properly, and if they are provided with the correct environment for flowering, the plants listed in this section will reward the indoor gardener with a more or less continuous production of flowers throughout their useful life. As is the case with any indoor plant, these plants will eventually become overgrown and will need repotting and pruning, as well as other renovative care.

Begonia × *hiemalis*

Rieger begonia, hiemalis begonia

This group of begonias was originally developed as a hybrid by crossing a bulbous winter-flowering begonia with hardy and vigorous tuberous begonias. Many cultivars of this begonia are currently popular; they are available in florists' shops, where they can be purchased in flower at any time of the year. They tend to flower more in the winter than at other seasons. Hiemalis begonias are low growing and very bushy. Many are pendulous and are attractive used in hanging baskets. Some of the newer cultivars have bronze or red foliage. The flowers are usually large and double and are available in yellows, reds, whites, and oranges. They prefer cooler locations than most other begonias, but they do not like drafts. Provide them with plenty of light during flowering.

Saintpaulia 'Mauna Loa' (striped flower)

LIGHT: Provide at least 4 hours of curtain-filtered sunlight in a bright south-, east-, or west-facing window.

WATER: Keep the plant evenly moist. Water thoroughly and discard drainage.

HUMIDITY: Average indoor humidity.

TEMPERATURES: 50° F to 55° F nights, 65° F to 70° F days.

PROBLEMS: This plant is subject to crown rot if planted deeply, watered over the crown, or watered late in the day. Some cultivars are subject to powdery mildew.

PROPAGATION: Take stem cuttings at any time.

GROOMING: Pinch out stem tips of young or regrowing plants to improve the plant's form. Do not destroy flower buds! Keep the plant to the desired height and shape with light pruning or clipping at any time.

REPOTTING: Cut back and repot when flowering stops.

FERTILIZATION: Fertilize all year, but more heavily in the summer.

Saintpaulia 'Dixie Moonbeam' (striped flower)

Begonia × *semperflorens-cultorum*

Wax begonia, fibrous-rooted begonia

These are the most popular of the fibrous-rooted begonias. Many cultivars and hybrids exist, and all are bushy plants with shiny (waxy), heart-shaped leaves. Given ample light they will bloom profusely in a variety of colors. Wax begonias are most commonly used as outdoor bedding annuals or in hanging baskets in patio gardens. Light fertilization, good light, and warmth are all these plants need to flourish indoors.

LIGHT: Provide at least 4 hours of curtain-filtered sunlight in a bright south-, east-, or west-facing window.

WATER: Keep the plant evenly moist. Water thoroughly and discard drainage.

HUMIDITY: This plant requires moist air. Use a humidifier for best results.

TEMPERATURES: 65° F to 70° F nights, 75° F to 80° F days.

PROBLEMS: This plant is subject to crown rot if planted deeply, watered over the crown, or watered late in the day. The leaves will scorch (turn brown and curl) if this plant is in a draft or in dry air; the stems will be spindly and weak if the light is too low. Mealybugs and powdery mildew disease can be a problem.

PROPAGATION: Take stem cuttings at any time. Seeds are also available, but are a bit difficult to grow.

GROOMING: Pinch out stem tips of young or regrowing plants to improve form, but be careful not to destroy flower buds in the process. Keep the plant at the desired height and shape with light pruning or clipping at any time.

REPOTTING: Cut back and repot when flowering stops.

FERTILIZATION: Fertilize three times a year, in spring, midsummer, and early fall.

Begonia × *hiemalis*

Begonia × *semperflorens*

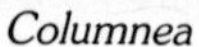

Columnea

Episcia cupreata

Columnea species
Columnea

If given enough light, these gesneriads will bloom throughout the year. There are over 150 species, with a variety of leaf textures and sizes and many flower colors in shades of yellow-orange or red. Columnea produce trailing stems with closely spaced opposite leaves. Flowers appear all along the stems. The blossoms are very large compared to the smaller, hairy foliage. As hanging baskets, they can get large if not pruned back from time to time. Be sure and keep these plants out of drafts of dry air, particularly in the winter.

LIGHT: Provide at least 4 hours of curtain-filtered sunlight in a bright south-, east-, or west-facing window.

WATER: Keep the plant evenly moist during growth and flowering. Allow it to dry between waterings at other times.

HUMIDITY: This plant requires moist air. Use a humidifier for best results.

TEMPERATURES: 55° F to 60° F nights, 70° F to 80° F days.

PROBLEMS: The leaves will scorch (turn brown and curl) if this plant is in a draft, in dry air, or exposed to too much sun. Poor drainage, excessive watering, or drainage water left in the saucer will cause root rot.

PROPAGATION: Take cuttings from recently matured stems or shoots.

GROOMING: Cut out oldest branches occasionally to encourage the formation of young branches. Most columnea do not branch on their own, but need to be cut back to encourage new branches.

REPOTTING: Infrequent repotting is best for this plant.

FERTILIZATION: Fertilize only during the late spring and summer months.

Episcia species
Flame violet

There are many flame violet cultivars, each with distinctive foliage texture and variegated coloring. Many indoor gardeners value them for this foliage alone. In the summer or in warm, humid conditions (such as those provided in terrariums), flame violets will produce small flowers in reds, oranges, pink, or yellow. The stems of these plants are trailing, which explains why flame violets are generally used in hanging baskets or as ground covers in well-lit terrariums. To keep the plants looking their best, grow them in a damp environment and cut back the spindly stems to induce branching. Various cultivars, planted in indoor planters with a fluorescent light approximately 1 foot above the plants, make a stunning conversation piece even if they are not blooming. Another common name for *Episcia* species is Persian carpet, which is very appropriate if they are grown to spread out on a table or windowsill. While most flame violets grow medium-large, there is a true miniature available called 'Silver Skies'.

LIGHT: Bright, indirect light.

WATER: Keep moist, not wet, when actively growing. Allow to dry between waterings at other times.

Exacum affine

HUMIDITY: This plant requires moist air. Use a humidifier for best results.

TEMPERATURES: 55° F to 65° F nights, 65° F to 75° F days.

PROBLEMS: The leaves will scorch (turn brown and curl) if the plant is in a draft, in dry air, or exposed to too much sun.

PROPAGATION: Runners can be rooted in soil to form new plantlets.

GROOMING: Pinch off the tips of the stems to encourage branching. Cut back the plant when it has stopped blooming to encourage new growth.

REPOTTING: Repot each year in the spring.

FERTILIZATION: Fertilize only during the late spring and summer months.

Impatiens wallerana 'Blitz'

Nematanthus 'Black Gold'

Exacum affine

Arabian violet, German violet

Arabian violets are popular because they will bloom as a small plant in a 4-inch pot. Plants are commonly covered with tiny blue flowers with yellow centers. Many florists carry them throughout the fall and winter. The seedlings must be handled carefully, because slight injuries may lead to stem rot and cankers. Keep the tiny seedlings in moist air and out of direct sun. As the plants get bigger, provide some direct sun in the fall to encourage blooming. Never place the plants in cool drafts or cold water.

LIGHT: Provide curtain-filtered sunlight in the summer in a south- or west-facing window. About 4 hours of direct sun are needed in the winter.

WATER: Keep the plant evenly moist. Water thoroughly and discard drainage.

HUMIDITY: This plant requires moist air. Use a humidifier for best results.

TEMPERATURES: 55° F to 60° F nights, 70° F to 75° F days.

PROBLEMS: This plant will die back if the roots are damaged from soil that is dry or high in soluble salts. Crown rot may occur if it is planted too deeply, watered over the crown, or watered late in the day. Whiteflies can be a problem.

PROPAGATION: Start from seeds in the spring. Begin in a small pot and transplant into larger pots as needed.

GROOMING: It is best to discard this plant after flowering.

REPOTTING: Transplant seedlings several times as the plants grow.

FERTILIZATION: Fertilize only during the late spring and summer months.

Impatiens wallerana

Patient Lucy, Busy Lizzie

This species is particularly suited for indoor gardening because it tolerates moderately lit sites and is relatively easy to grow. Cultivars with variegated leaves are available. Flowers are produced throughout the year, but it is best to pinch out the stems on new or regrowing plants to obtain better form before allowing the plant to bloom. If you bring Patient Lucys in from the patio garden in the fall, be sure they are free of pests, particularly whiteflies or spider mites.

LIGHT: Provide at least 4 hours of curtain-filtered sunlight in a bright south-, east-, or west-facing window.

WATER: Keep the plant evenly moist. Water thoroughly and discard drainage.

HUMIDITY: Average indoor humidity levels are satisfactory.

TEMPERATURES: 50° F to 55° F nights, 60° F to 65° F days.

PROBLEMS: Spider mites often attack this plant, especially if it is too dry. The leaves will drop if soil moisture is too wet or too dry. The plant will not bloom if light levels are too low.

PROPAGATION: Start from seeds. Begin in a small pot and transplant into larger pots, as needed.

GROOMING: It is best to discard this plant after flowering.

REPOTTING: Transplant seedlings several times as the plants grow.

FERTILIZATION: Fertilize all year, but more heavily in the summer.

Nematanthus species

Nematanthus

This gesneriad produces flowers along the length of trailing stems. The family is the only one (other than African violets) whose hybrids and even some species are truly everblooming under optimum growing conditions. The flowers of the small-leaved cultivars resemble a candycorn or tiny fish. The leaves are small and exceptionally shiny, and are produced closely along the stems. The plant can be shaped into a semi-upright form or grown in a small hanging basket. *N. wettsteinii*, an everblooming species, is a parent plant for some of the beautiful new hybrids. All have stunning, polished-looking foliage. The flowers are usually smaller in size than the foliage and their yellow to orange colors contrast beautifully with the dark, sometimes nearly black, foliage. One of the widely grown large cultivars, 'Tropicana', has large, very dark, shiny foliage with relatively small, orange-striped, darker guppy-type blossoms. 'Tropicana' grows best in an 8- or 10-inch basket; the mature plant is a showpiece that may grow to 3 feet across.

LIGHT: Provide 6 or more hours of curtain-filtered sunlight in a bright south-, east-, or west-facing window.

WATER: Keep the plant evenly moist. Water thoroughly and discard drainage.

HUMIDITY: Average indoor humidity levels are satisfactory, but regular misting improves production of blossoms.

TEMPERATURES: 55° F to 60° F nights, 70° F to 80° F days.

Aeschynanthus 'Flash'

Agapanthus 'Peter Pan White'

PROBLEMS: The plant will not bloom if light levels and humidity are too low, and will die back if the roots are damaged from dry soil or overwatering. Spider mites often attack this plant, especially if it is too dry.

PROPAGATION: Take stem cuttings at any time.

GROOMING: Prune back well in early spring. Cut out older, woody stems to improve form and encourage branching, but do not destroy flower buds!

REPOTTING: Infrequent repotting is best. Repot in the winter or early spring, if needed.

FERTILIZATION: Fertilize lightly all year around.

PLANTS THAT BLOOM SEASONALLY OR INTERMITTENTLY INDOORS

Many would say that the most "natural" way plants bloom is seasonally. Although in some respects an indoor garden is without seasons, indoor plants still may react to the subtle light, temperature, or humidity changes that occur yearly, even indoors. Thus they tend to bloom seasonally, or at least from time to time, rather than continuously.

A skillful indoor gardener can have colorful blooms constantly in the garden. Select the proper mix of plants from the following pages. In most cases, the time of year in which the plant flowers is given in the descriptive paragraph.

***Aeschynanthus* species**
Lipstick plant, basket plant

This gesneriad has thick reddish foliage abundantly produced on trailing stems. It is best used in a hanging basket. The tube-like flowers are borne mostly on the ends of the branches in a cluster of bright red or orange blooms, each of which resembles a stick of bright lipstick when young. One widely grown species, *A. marmoratus*, has reddish-splashed foliage, with sparse greenish blossoms. It is commonly called zebra vine. In good light lipstick vines can be covered with hundreds of blooms. These plants require night warmth and good winter light to do well. They can often be purchased from florists. Small plants will take some time to grow into a large basket.

LIGHT: Provide 4 or more hours of curtain-filtered sunlight in a south-facing window or a bright east- or west-facing window.

WATER: Keep the plant evenly moist. Water thoroughly and discard drainage.

HUMIDITY: Average indoor humidity levels will be satisfactory; however, regular misting encourages blooming.

TEMPERATURES: 65° F to 70° F nights, 75° F to 85° F days.

PROBLEMS: The plant will not bloom if light levels are too low, and will die back if the roots are damaged from soil that is dry or high in soluble salts. Mealybugs often attack this plant.

PROPAGATION: Take cuttings from stems or shoots before they have hardened or matured.

GROOMING: Prune back after flowering or fruiting. Start new plants and replace older specimens when they get weak or woody.

REPOTTING: Infrequent repotting is best for this plant.

FERTILIZATION: Fertilize lightly each month.

***Agapanthus* species**
Lily-of-the-Nile

Lily-of-the-Nile plants are large and bear clusters of blue or white flowers on tall stalks in the summer. The plants bloom better when allowed to mature and get slightly pot-bound. Of course, this means you must be careful not to let them get too dry between waterings if they are to mature properly. *Agapanthus orientalis* is the species with the largest plants, occasionally reaching 5 feet tall with bloom clusters that sometimes contain up to 100 flowers. Plants of the other species are around 2 feet tall.

LIGHT: Provide curtain-filtered sunlight in the summer, in a south- or west-facing window. Keep in about 4 hours of direct sun in the winter.

WATER: Keep the plant evenly moist. Water thoroughly and discard drainage.

HUMIDITY: Average indoor humidity levels are adequate.

TEMPERATURES: 50° F to 55° F nights, 65° F to 70° F days.

PROBLEMS: Leaf yellowing will occur in low light or when the soil is too wet or too dry.

PROPAGATION: New plants are started by dividing older specimens. Seeds are also available but are generally harder to grow.

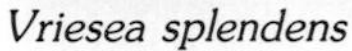

Vriesea splendens

Guzmania lingulata

GROOMING: Pick off yellowed leaves as they occur.

REPOTTING: Infrequent repotting is best for this plant.

FERTILIZATION: Fertilize only when growing actively or flowering.

Begonia × *cheimantha*
Christmas begonia, Lorraine begonia

These hybrid begonias are becoming more popular because they are profuse winter bloomers. They are compact, dwarf bushy plants and are most frequently used in hanging baskets because of the tendency for the stems to arch outward. Given enough light, Christmas begonias will be covered with pink or white single flowers on long stems or racemes. Since they are winter bloomers, they must be given special care to make sure they receive enough light and warmth and are kept evenly moist while they are flowering. After flowering, the plants become semidormant until late spring. Keep them drier during this period.

LIGHT: Provide at least 4 hours of curtain-filtered sunlight in a bright south-, east-, or west-facing window.

WATER: Keep the plant very moist during growth and flowering. Allow it to dry between waterings at other times.

HUMIDITY: Average indoor humidity levels are satisfactory.

TEMPERATURES: 65° F to 70° F nights, 75° F to 80° F days.

PROBLEMS: This plant is subject to crown rot if planted too deeply, watered over the crown, or watered late in the day. The leaves will drop if soil is too wet or too dry. Some cultivars are subject to powdery mildew.

PROPAGATION: Take stem cuttings at any time.

GROOMING: Prune back after flowering or fruiting.

REPOTTING: Cut back and repot when flowering stops.

FERTILIZATION: Fertilize only when growing actively or flowering.

Bromeliads—Many Genera and Species

The bromeliads comprise a large plant family, most of which are native to the American tropics. Most bromeliads are epiphytes, and grow in the forests on tree limbs and rocks. The roots are used mainly for attachment. Nutrients and water are taken up through the leaves. Many plants are urn- or vase-shaped and will hold water at the base of the rosetted leaves. Plant debris that falls into the water serves to provide the needed nutrition.

Bromeliads have recently become very popular as "exotic" plants for use in homes and offices. Many are relatively easy to maintain indoors grown in pots in a light, well-drained potting soil. Some bromeliads are grown for their flowers, and others are grown primarily for their foliage. The foliage and flowers of the different genera vary in size, color, and variegation.

As flowering plants, they are best when purchased in bloom with the flowers one-quarter opened. Flowers commonly last several months. Bromeliads need lots of sun and warmth to bloom. Dedicated indoor gardeners can induce the plants to bloom, often by placing them in a plastic bag with a ripe apple for a few days— the ethylene gas given off by the apple helps the plant initiate flower buds.

Bromeliad plants only bloom once. After flowering, new plants appear either from the side or center of the old plant. For centrally sprouting plants, remove old leaves as they turn brown. Cut off side growths along with some roots and start new plants. In pot culture, the foliage of bromeliads does not need to be watered. In fact, such practice may encourage rotting.

LIGHT: Give the plant 4 hours or more of direct sunlight in a south-facing window.

WATER: Keep the plant evenly moist. Water thoroughly and discard drainage.

HUMIDITY: Dry air is generally not harmful for this plant, but keep it out of drafts.

TEMPERATURES: 65° F to 70° F nights, 75° F to 80° F days.

PROBLEMS: Poor drainage, excessive watering, or water left in the saucer causes root rot. Bromeliads are subject to infestations of scale and mealybugs.

Guzmania × 'Cherry'

Aechmea fasciata

Ananas

PROPAGATION: Remove rooted side shoots or repot a new, centrally growing plant.

GROOMING: Pick off yellowed leaves as they occur.

REPOTTING: Infrequent repotting is best for this plant. Repot new plants.

FERTILIZATION: Fertilize only when growing actively or flowering.

Aechmea species
Living vase plant

The upright rosette of thick, silver-banded leaves distinguishes the striking *Aechmea chantinii.* The flowers last for several months. The most common of this group is *A. fasciata*, the urn plant. Its broad, thick leaves are mottled with stripes of gray and deep sea green. The conical rosette of pink bracts and large blue flowers provide a splendid effect. *A. fulgens* var. *discolor*, commonly known as the coralberry, features broad leaves that are green on top and purple underneath. The contrast in the foliage is heightened by the purple flower. Red berries form after the flower dies.

Ananas species
Pineapple

Pineapples are the fruit of *Ananas comosus*. You can grow one by simply cutting off a bit of the fruit along with the fruit's tuft of leaves, planting it in soil (it's a terrestrial bromeliad), and placing it in full sun. Narrow, gray-green leaves with prickly ribbing running up the sides will grow to form a striking rosette; the pineapple fruit grows from the center in an unusual display, but this only happens after several years. *A. comosus* 'Variegatus', commonly known as the ivory pineapple, is thought to have much more attractive foliage.

Billbergia species
Vase plant

These are among the easiest bromeliads to grow, but they flower for only a short time. *Billbergia nutans*, known as queen's tears, has grasslike gray-green leaves. Amidst the foliage is an arching spray of pink and green flowers. Another, *B. pyramidalis,* sports long, green, straplike leaves. The ornamental inflorescence has bright red bracts tipped with violet and upright scarlet flowers with yellow stamens.

Guzmania species

Guzmanias are one of the familiar vase-shaped bromeliads that can get to be 20 inches across. They bloom from late winter to summer, depending on the species. The true flowers are small, but are surrounded by large, showy bracts in reds, yellows, or oranges. Keep the plants moist by keeping water in the vase-shaped rosette of leaves when the plants are growing and flowering.

Tillandsia species
Blue torch, Spanish moss

Many species of *Tillandsia* are available to indoor gardeners. Even the Spanish moss commonly seen as an epiphyte in the southern states is occasionally grown indoors. Most blue torches have narrow, arching leaves that are sometimes grasslike. Some of the smaller species are popular as hanging plants or dish-garden plants. The flower spikes are often swordlike and form on long stems, usually in the summer.

Vriesea species

This genus features many plants attractive for both their foliage and flowers. *Vriesea splendens*, a popular species, forms a rosette of wide, purple-banded leaves. The common name, flaming sword, refers to its flower cluster, a long spike of red bracts that surround the yellow flowers. The flowering spikes last for several weeks.

Daphne odora
Winter daphne

This winter-flowering plant produces clusters of small pink or reddish flowers with a very sweet fragrance. The leathery leaves on this woody plant are a shiny green, usually about 3 inches long. There is a variegated form available with creamy white leaf edges. The plants will get large, but are most attractive if kept pruned to 1 foot or so.

LIGHT: Provide at least 4 hours of curtain-filtered sunlight in a bright south-, east-, or west-facing window.

WATER: Approach dryness before watering, but water thoroughly and discard drainage each time.

HUMIDITY: Average indoor humidity levels are satisfactory.

TEMPERATURES: 40° F to 45° F nights, 60° F to 65° F days.

Billbergia pyramidalis

Tillandsia cyanea

Hoya

PROBLEMS: Poor drainage, excessive watering frequency, or water left in the saucer causes root rot. Leaves may get yellow from soil high in soluble salts, high pH, or lack of iron.

PROPAGATION: Take cuttings from recently matured stems or shoots.

GROOMING: Prune back well in early spring. Keep the plant to the desired height and shape with light pruning or clipping at any time.

REPOTTING: Repot in the winter or early spring as needed.

FERTILIZATION: Fertilize lightly once a year in early spring. Use an acid balanced fertilizer with trace elements.

***Hoya* species**
Wax plant

Hoyas are vining plants with thickened leaves. If given enough light, they produce clusters of extremely fragrant, waxy-looking flowers. Depending on the cultivar, wax plants bloom in the summer or fall. The flowers form on spurs, so be careful not to prune off these leafless extensions of the stems. Train the plants on a trellis or use in a hanging basket. Wax plant vines get quite long, but you can double them back to give a denser appearance to the plant. Many variegated and colored leaf forms are available.

LIGHT: Place in bright indirect light in a south-, east-, or west-facing window.

WATER: Keep the plant very moist during growth and flowering. Allow it to dry between waterings at other times.

HUMIDITY: Average indoor humidity levels are satisfactory.

TEMPERATURES: 50° F to 55° F nights, 60° F to 65° F days.

PROBLEMS: The plant will not bloom if light levels are too low.

PROPAGATION: Take stem cuttings at any time.

GROOMING: Keep the plant to the desired height and shape with light pruning or clipping at any time.

REPOTTING: Infrequent repotting is best for this plant. New plants must grow in a medium to large pot for some time before they will bloom or set fruit.

FERTILIZATION: Fertilize only when growing actively or flowering.

Orchids—Many Genera

Orchids are probably the most varied group of flowering plants. One or more genera are found throughout the world, in deserts, tropics, and Arctic regions. Orchids can be large or small, epiphytic or terrestrial, with fragrant or odorless blooms in many colors.

Although they have a reputation as being "exotic" plants, many species of orchids grow quite well indoors and require less routine care than many of the flowering houseplants included in this book. They are becoming widely available because improved methods of tissue culture propagation have significantly lowered their production costs.

When choosing an orchid, it is best to select a mature plant already in flower. Most orchids have rather unattractive foliage, and many may not bloom until they are several years old. Culture the terrestrial types in a well-aerated potting mix rich in peat. Purchase a fir bark mix for the epiphytic types. Summer heat may stress orchids indoors. This can be avoided by moving them outdoors or by gently moving air over them with a fan.

LIGHT: Provide curtain-filtered sunlight in the summer in a south- or west-facing window. In the winter they need about 4 hours of direct sun each day.

WATER: Keep the plant very moist during growth and flowering. Allow it to dry between waterings at other times.

HUMIDITY: This plant requires moist air. Use a humidifier for best results.

TEMPERATURES: Keep temperatures reasonably constant around 50° F to 60° F.

PROBLEMS: This plant will die back if the roots are damaged from soil that is dry or high in soluble salts. Poor drainage, excessive watering frequency, or water left in the saucer will cause root rot.

Dendrobium

Cattleya

PROPAGATION: Use mature plants—new plants are started by dividing an older specimen.

GROOMING: Pick off yellowed leaves as they occur.

REPOTTING: Infrequent repotting is best for this plant.

FERTILIZATION: Fertilize three times a year, in spring, midsummer, and early fall. Use an acid-balanced fertilizer and add trace elements once in the spring.

Brassia species
Spider orchid
Spider orchids bear flowers with long, narrow sepals that give rise to their common name. The plants are fairly large, with 15-inch flower spikes and leaves up to 10 inches long. They generally bloom in the fall or winter if given enough light.

Cattleya species
Cattleya, the large classical orchid most often seen in corsages, is good for beginners to try. The vigorous plants produce gorgeous blooms when they receive plenty of sun and average daytime temperatures.

Cymbidium species
The miniature cymbidiums are especially well suited for many indoor gardeners. Even with miniatures, however, the narrow, arching leaves must be given room. Give the plants some sun and cool nights to promote flowering, usually in late summer or fall. Many hybrids are available in a wide variety of flower colors. The flowers are long lasting, even as cut flowers in arrangements.

Odontoglossum

Brassia maculata

Dendrobium species
Dendrobiums are mostly epiphytic orchids, with both evergreen and deciduous plants available. Large flowers bloom in clusters or in a row along the stem. They last for at least one week and up to several months, depending on the species. These plants need plenty of sun, as do most orchids.

Odontoglossum species
Lily-of-the-valley orchid
These orchids need moist air and stable growing conditions. They are best suited for greenhouses, where they can get direct winter sun and curtain-filtered light in the summer. Many species and hybrids are available, most of which bear large, fragrant, and long-lasting flowers. Many will bloom twice a year.

Oncidium species
Butterfly orchid
The butterfly orchids are a large group of epiphytic orchids (they do not grow in soil). They produce stalks of yellow flowers speckled with brown. Flower size depends on the species.

Phalaenopsis species
Moth orchid
Phalaenopsis, commonly known as moth orchids, unfurl sprays of 2- to 3-inch flowers in a range of colors. The plant grows up to 30 inches high. These shade-loving plants are easy to grow at temperatures of 75° F during the day and 60° F at night.

Pelargonium species
Geranium
Geraniums thrive indoors, provided you have cool spots where the light is ample. A minimally heated sun room or window box is ideal. Many cultivars get quite large, and the dwarf or miniature cultivars are often preferred by indoor gardeners. Although they may flower all year, geraniums make particularly good winter bloomers indoors. Do not overwater or overfertilize these plants, and keep them pruned back. Discard them when they get too leggy and weak.

Cymbidium

Phalaenopsis

Oncidium

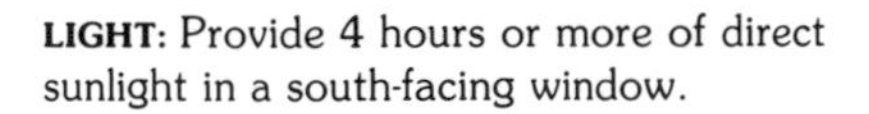

LIGHT: Provide 4 hours or more of direct sunlight in a south-facing window.

WATER: Water thoroughly, but allow the soil to dry out between waterings.

HUMIDITY: Average indoor humidity levels are satisfactory.

TEMPERATURES: 50° F to 55° F nights, 60° F to 65° F days.

PROBLEMS: Spider mites often attack this plant, especially if it is too dry. Poor drainage, excessive watering, or water left in the saucer causes root rot.

PROPAGATION: Take cuttings from stems or shoots before they have hardened or matured.

GROOMING: Prune back well in early spring.

REPOTTING: Repotting should be done each year.

FERTILIZATION: Fertilize only during the late spring and summer months.

Schlumbergera species

Thanksgiving cactus, Christmas cactus, Easter cactus

These plants have branches made up of smooth, flat, 1½-inch-long segments or joints that arch outward. Flowers are borne on the ends of these stems at the appropriate time of the year (depending on the cultivar chosen). The flowers are 2 or 3 inches long, in oranges or reds. Budding results from short days during October and November, accompanied by a cold shock. Many gardeners achieve this by placing the plants outdoors for a time during frost-free days and nights in the fall.

Pelargonium peltatum 'Galilee'

Schlumbergera truncata

LIGHT: Place the plant in a bright, indirectly lit south-, east-, or west-facing window.

WATER: Keep the plant very moist during growth and flowering. Allow it to dry between waterings at other times.

HUMIDITY: Average indoor humidity levels are satisfactory.

TEMPERATURES: 40° F to 45° F nights, 60° F to 65° F days to set flower buds; 50° F to 55° F nights, 65° F to 70° F days at other times.

PROBLEMS: This plant will die back if the roots are damaged from soil that is dry or high in soluble salts.

PROPAGATION: Take cuttings from recently matured stems or shoots when the plant is not in flower.

GROOMING: Prune back after flowering or fruiting, if needed.

REPOTTING: Infrequent repotting is best for this plant.

FERTILIZATION: Fertilize only when growing actively or flowering.

Spathiphyllum 'Mauna Loa'

Streptocarpus

Spathiphyllum **species**
Peace lily, spathe flower

These easy-to-maintain plants are unusual in that they bloom even under moderately lighted conditions. The fragrant "flower" is a white bract, the spathe, which surrounds the spadix or "candle" of tiny true flowers. Peace lilies bloom intermittently throughout the year, but more commonly in the spring. When not in flower, they are still valued for their foliage. Some cultivars, such as 'Mauna Loa', reach 3 feet tall or higher.

LIGHT: This plant survives in low light, light by which you can read comfortably. Bright, indirect light is best for flowering.

WATER: Keep the plant evenly moist. Water thoroughly and discard drainage.

HUMIDITY: Average indoor humidity levels are satisfactory.

TEMPERATURES: 55° F to 60° F nights, 70° F to 75° F days.

PROBLEMS: The leaves will scorch (turn brown and curl) if this plant is in a draft or in dry air.

PROPAGATION: New plants are started by dividing an older specimen. Seeds are also available, but are generally harder to grow.

GROOMING: Pick off yellowed leaves as they occur.

REPOTTING: Infrequent repotting is best for this plant. Repot in the winter or early spring, as needed.

FERTILIZATION: Fertilize only during the late spring and summer months.

Streptocarpus **species**
Cape primrose

These gesneriads are native to the southern tip of Africa, thus the common name Cape primrose. If given enough light they bloom at any time of year as they reach maturity. Many cultivars are available, some of which get quite large. Flowers, in pinks, blues, or white, as well as multicolored, form on tall stalks above rosettes of leaves. Mature leaves are commonly 8 to 12 inches long. In Scandinavia Cape primrose is grown in sheltered window boxes outside, hence another common name, Swedish gloxinia (the blossoms have a slight resemblance to florist's gloxinia). The hybridizers are now working on a miniature species of *Streptocarpus* suitable for 3-inch pots. They are only 6 inches tall at maturity, but perform just as well as their larger cousins. 'Snow Drop', an all-white little beauty, is already widely available, and other assorted colors are being grown by specialty growers.

LIGHT: Provide at least 4 hours of curtain-filtered sunlight, in a bright south-, east-, or west-facing window.

WATER: Keep the plant very moist during growth and flowering. Allow it to dry between waterings at other times.

HUMIDITY: This plant requires moist air. Use a humidifier for best results.

TEMPERATURES: 65° F to 70° F nights, 75° F to 80° F days.

PROBLEMS: The plant will not bloom if light levels are too low. The leaves will scorch (turn brown and curl) if this plant is in a draft or in dry air. Spider mites often attack this plant, especially if it is too dry.

PROPAGATION: Start from seeds. Begin in a small pot and transplant to larger pots, as needed. Leaf cuttings can also be made with some cultivars.

GROOMING: Pick off yellowed leaves as they occur. Cut flower stalks if you wish.

REPOTTING: Repot when new growth starts after dormancy.

FERTILIZATION: Fertilize only when growing actively or flowering.

PLANTS THAT NEED A REST AFTER BLOOMING

Most plants that have evolved as perennials in climates where there are great seasonal changes in dryness, heat, or cold have genetic dormancy or semidormancy properties. You must give some of these plants a rest after blooming, in spite of the fact that the plant may actually be growing in the relatively "seasonless" indoor environment. Gardeners who do not know of this need may wonder why the plant seems to be dying or failing to perform as expected after a bloom period.

These plants come from many unrelated genera. Most bulbous plants need a complete dormancy rest, and are thus grouped together following a generalized discussion concerning their care. Some of the other plants in this grouping need only partial dormancy—a period where watering frequency is lessened and fertilization is stopped. Such a period may last from one to three months. Many indoor gardeners find it easiest to discard these plants after their flowering period. Of course, some of the bulbous plants can be planted outdoors if they are winter hardy in your area of the country.

Abutilon 'Moonchimes'

Abutilon species

Flowering maple, bellflower

This treelike plant with maple-shaped leaves grows quickly and gets quite large. The leaves, often dappled with yellow or white, are attractive by themselves. The large pendulous flowers, in shades of yellow, orange, or red, are borne in the spring if they are given ample light. Keep the plant constantly pruned to maintain its size and shape. Abutilons benefit from as much light as they can get, but are subject to leaf scorch if allowed to dry out between waterings.

LIGHT: Provide curtain-filtered sunlight in the summer in a south- or west-facing window. Give about 4 hours of direct sun in the winter.

WATER: Keep the plant evenly moist. Water thoroughly and discard drainage.

HUMIDITY: This plant requires moist air. Use a humidifier for best results.

TEMPERATURES: 50° F to 55° F nights, 65° F to 70° F days.

PROBLEMS: Spider mites often attack this plant, especially if it is too dry. The leaves will scorch (turn brown and curl) if this plant is in a draft or in dry air. Never allow the plant to dry out.

PROPAGATION: Take cuttings from recently matured stems or shoots.

GROOMING: Keep the plant to the desired height and shape with light pruning or clipping at any time. Give these plants plenty of room.

Achimenes 'Minuette'

REPOTTING: Infrequent repotting is best for this plant.

FERTILIZATION: Fertilize all year, but more heavily in the summer.

Achimenes species

Rainbow flower, magic flower

The many cultivars of *Achimenes* species offer a variety of flower colors from light blue to deep red and yellow. They are often found in florists' shops. Like their African violet relatives, they need warmth to grow well and ample light to blossom. The genus *Achimenes* contains bushy plants that are often trained into hanging baskets. The foliage of many of the cultivars is attractive by itself, especially if the branches are properly pinched and trained when the plants are young.

LIGHT: Provide at least 4 hours of curtain-filtered sunlight in a bright south-, east-, or west-facing window.

WATER: Keep the plant evenly moist. Water thoroughly and discard drainage.

HUMIDITY: This plant requires moist air. Use a humidifier for best results.

Begonia × *tuberhybrida* 'Nonstop Orange'

TEMPERATURES: 65° F to 70° F nights, 75° F to 80° F days.

PROBLEMS: The plant will not bloom if light levels are too low, and the leaves will drop if soil moisture is too wet or too dry. This plant will die back if the roots are damaged from soil that is dry or high in soluble salts.

PROPAGATION: New plants are started by dividing an older specimen. Seeds can be planted in spring.

GROOMING: Pinch out stem tips of young or regrowing plants to improve form, but do not destroy flower buds.

REPOTTING: Let the plant die back after flowering, then remove rhizomes and repot. Pack the rhizomes in dry peat moss or vermiculite, and store at 60° F. Keep recently potted plants warm and only moderately moist.

FERTILIZATION: Fertilize three times a year, in spring, midsummer, and early fall.

Begonia × *tuberhybrida*

Tuberous begonia

Cultivars of tuberous begonias produce the largest flowers of all types of begonias used in indoor gardening. Most are large plants that need good light, cool temperatures, and moist soil and air. It is best to buy a mature tuber, plant it, enjoy a flowering period, and then discard it or place it in the garden. Older plants tend to get spindly and weak indoors.

× *Amarcrinum memoria-corsii*

LIGHT: Provide curtain-filtered sunlight in the summer in a south- or west-facing window. Give about 4 hours of direct sun in the winter.

WATER: Keep the plant very moist at all times, but do not allow to stand in water.

HUMIDITY: This plant requires moist air. Use a humidifier for best results.

TEMPERATURES: 50° F to 55° F nights, 60° F to 65° F days.

PROBLEMS: The leaves will scorch (turn brown and curl) if this plant is placed in a draft or in dry air. This plant is subject to crown rot if planted deeply, watered over the crown, or watered late in the day.

PROPAGATION: Start new plants from the small tubers that develop beside the parent.

GROOMING: It is best to discard this plant after flowering.

REPOTTING: Not usually done.

FERTILIZATION: Do not fertilize when in flower. Fertilize lightly at other times.

Bulbous Plants for Indoor Flowering

One rewarding aspect of indoor gardening is that you can persuade plants to bloom out of season. By duplicating—but shortening—the stages bulbs go through in your garden, you can have tulips, daffodils, hyacinths, and other bulbous plants blooming in the house while the wind drifts snow outdoors.

Some of the bulbous plants included in this section do not need to be given a "winter" treatment. They can be purchased from the garden center and immediately cultured indoors for growth and flowering. Most of these plants do need to rest or undergo a dormant period after their initial flowering period. It is for this reason that many indoor gardeners prefer to discard such plants after they have finished blooming.

Certain signs indicate that the plant is ready to rest. In spite of correct watering and light, the foliage will begin to yellow and wilt after blooming has finished. When this happens, water less frequently and do not fertilize further. Do not stop watering immediately. In a few weeks new growth may start, or you can begin watering the plant again to encourage new growth.

Forcing Bulbs Indoors

It's easy to force tulips, daffodils, hyacinths, and the little bulbs—crocuses and grape hyacinths—to bloom indoors ahead of their normal time outdoors. Grow them as Christmas gifts for friends. Buy the largest bulbs you can find of the type you wish to grow. Read the catalogs carefully to select those recommended for forcing. Most retail nurseries and garden centers will carry bulbs for forcing in the fall. For those bulbs you can't find in your local store, use mail order catalogs. Order by September so the bulbs will be delivered in early fall. Then follow these steps:

Prepare a growing medium of equal parts soil, builder's sand, and peat moss. To each 5-inch pot of this mix, add 1 teaspoon of bone meal. If you don't want the bother of mixing soil, buy all-purpose medium.

Pot size depends on the type and quantity of bulbs. One large daffodil or tulip bulb can be planted in a 4- or 5-inch pot in which three crocuses or other smaller bulbs would otherwise fit. For six tulips, daffodils, or hyacinths, you will need an 8- to 10-inch pot. When you plant these large bulbs, cover the tops of tulips and hyacinths with 1 inch of soil; however, do not cover the necks and tops of daffodils. Cover the smaller bulbs, like crocuses, with 1 inch of soil, then water thoroughly.

Many bulbs need a period of coolness after potting so that they can form a vigorous root system. Without a potful of roots, they cannot bloom prolifically later on. (It is also possible to purchase preplanted containers of bulbs conditioned to begin the forcing process.) Tradition has it that you bury these pots of bulbs in a bed of cinders outdoors in a coldframe, leaving them there until at least New Year's Day (except some bulbs conditioned to bloom for Christmas). This system is impractical for most of us today, and there are easier ways to accomplish the same thing.

Find a cool, frost-free place where the bulbs can be forced. A garage that is attached to the house, but not heated, is a good place for bulbs to form roots. A cool attic or basement will also do. A temperature range of 35° F to 55° F will promote root growth. Keep the soil evenly moist throughout this period.

You can start forcing the bulbs when sprouts begin to push up through the soil, usually after January 1. Bring the pots indoors, a few each week so you will have blooms over a longer period of time, and place them in a sunny, cool (55° F to 70° F) spot. Never allow the soil to dry out. The cooler the air, the longer the flowers will last. Keep bulbs away from sources of heat such as radiators and gas heaters. Bring all pots into a warm and sunny location by late February.

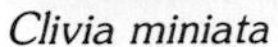

Clivia miniata

Crinum 'Cape Dawn'

Problems in forcing bulbs are few, but here are some that might occur:

- Flower buds of forced bulbs may blast (fail to open) if the soil is allowed to dry out severely after they begin to grow.
- Sometimes bulbs get basal rot; this is seldom your fault. If foliage suddenly turns yellow and stops growing, give it a gentle tug. Chances are you'll find it loose in the pot, and a rootless rotted bulb in the soil.

After the flowers fade, keep the foliage in good health by providing moisture and sunlight. As soon as any danger of hard freezing is past, move the bulbs to an out-of-the-way place outdoors where the foliage can continue to mature and store up strength for another year's blooms. Although they will not stand forcing a second year, you will find them useful additions for the outdoor garden. Plant the bulbs when you bring them outside, or leave them in pots until the following autumn, transferring them then to the open ground. Store the bulbs in a cool, dark, ventilated place if you need the pots before autumn.

× *Amarcrinum memoria-corsii*
Crinodonna

This large bulbous plant is reasonably easy to grow once it has become adapted and stabilized in its environment. It does need ample light to produce the large fragrant pink flowers. These flowers form in a cluster at the tip of a 3-foot stalk in late summer. The leaves can be 2 feet long, so give it plenty of room. As with most bulbs, this plant needs a dormancy rest. Provide this over the winter.

LIGHT: Give the plant 4 hours or more of direct sunlight in a south-facing window.

WATER: Keep the plant very moist during growth and flowering. Allow it to dry between waterings at other times.

HUMIDITY: Average indoor humidity levels are satisfactory.

TEMPERATURES: 50° F to 55° F nights, 65° F to 70° F days.

PROBLEMS: It will not bloom if roots are disturbed.

PROPAGATION: Start new plants from the small bulblets that develop beside the parent.

GROOMING: Give these plants plenty of room. Remove old leaves as the plant goes dormant.

REPOTTING: Infrequent repotting is best for this plant—once every three or four years. Leave the top third of the bulb out of the soil when replanting.

FERTILIZATION: Fertilize only when growing actively or flowering.

Clivia miniata
Kaffir lily

This rather large plant has long, strap-like leaves with large clusters of orange and yellow flowers borne on foot-long stems. Many hybrids bloom well in the late winter when given good indirect light. The plant needs cool temperatures and goes dormant during the fall. It is not a true bulbous plant, but closely resembles one in habit and form.

LIGHT: Provide at least 4 hours of curtain-filtered sunlight in a bright south-, east-, or west-facing window.

WATER: Keep the plant very moist during growth and flowering. Allow it to dry between waterings at other times. Withhold water from September to January.

HUMIDITY: Average indoor humidity levels are satisfactory.

TEMPERATURES: 50° F to 55° F nights, 60° F to 65° F days.

PROBLEMS: Poor drainage, excessive watering, or water left in the saucer causes root rot.

PROPAGATION: New plants are started by dividing an older specimen.

GROOMING: Remove old leaves as the plant goes dormant.

REPOTTING: Infrequent repotting is best for this plant.

FERTILIZATION: Fertilize only when growing actively or flowering.

Crinum species
Bengal lily, milk-and-wine lily

Bengal lilies are one of the largest of the bulbous plants used for indoor growing and flowering. The pink, red, and white flowers are fragrant and sometimes 6 inches across. They are borne in clusters on top of a 3-foot stalk. The leaves of the bulb are narrow and 4 feet long. It is a magnificent plant, but does need a lot of room. The plant flowers in the fall, so keep it well lit, moist, and fertilized from spring until October. A moderately dry resting period over winter is all that is needed for Bengal lilies.

Crocus

Eucharis grandiflora

LIGHT: Provide curtain-filtered sunlight in the summer in a south- or west-facing window.

WATER: Keep the plant very moist during growth and flowering. Allow it to dry between waterings at other times.

HUMIDITY: Average indoor humidity levels are satisfactory.

TEMPERATURES: 50° F to 55° F nights, 65° F to 70° F days.

PROBLEMS: No significant problems.

PROPAGATION: Start new plants from the small bulblets that develop beside the parent.

GROOMING: Remove old leaves as the plant goes dormant. Cut flower stalks if you wish. Give the plants plenty of room.

REPOTTING: Infrequent repotting is best for this plant.

FERTILIZATION: Fertilize only when growing actively or flowering.

Crocus species
Crocus

Crocus, because of their small size, are well suited for a midwinter flowering pot plant. Florists often sell them in late winter. The plants generally get only a few inches high and they look best planted in clumps in a broad, shallow pot. Varieties are available in many colors and shades. The corms must be given a cold (35° F) treatment in the pot prior to forcing. Many indoor gardeners pot up the newly purchased mature corms in October, place them in a cool (not freezing) spot until January, and bring them inside for forcing. After the foliage has died back, either place the plants in the garden or discard them.

LIGHT: Place the plant in a bright, indirectly lit south-, east-, or west-facing window.

WATER: Keep the plant evenly moist. Water thoroughly and discard drainage.

HUMIDITY: Average indoor humidity levels are satisfactory.

TEMPERATURES: 40° F to 45° F nights, 60° F to 65° F days.

PROBLEMS: No significant problems.

PROPAGATION: Start new plants from the small offsets that develop beside the parent corm. It takes several years to get a new corm up to blooming size.

GROOMING: Cut flower stalks if you wish. It is best to discard this plant after flowering.

REPOTTING: None needed.

FERTILIZATION: Do not fertilize when in flower. Fertilize lightly at other times.

Eucharis grandiflora
Amazon lily

Amazon lilies are easy-to-grow bulbs that flower even in indirect light indoors. They are very large plants. The flowers are showy, white, and borne in groups of three to six at the tip of a 2-foot stalk. The bulb may bloom as often as three times a year. The plant must be kept very moist and must be well fertilized while growing. Only a mild resting period is needed. Never let it get excessively dry, and do not fertilize it when it is not growing. Keep the plant warm, especially at night.

LIGHT: Place the plant in a bright, indirectly lit south-, east-, or west-facing window.

WATER: Keep the plant very moist during growth and flowering. Allow it to dry between waterings at other times.

HUMIDITY: This plant requires moist air. Use a humidifier for best results.

TEMPERATURES: 65° F to 70° F nights, 75° F to 80° F days.

PROBLEMS: The leaves will scorch (turn brown and curl) if this plant is placed in a draft or in dry air. Poor drainage, excessive watering, or water left in the saucer causes root rot.

PROPAGATION: Start new plants from the small bulblets that develop beside the parent.

GROOMING: Give these plants plenty of room.

REPOTTING: Infrequent repotting is best for this plant.

FERTILIZATION: Fertilize only when growing actively or flowering.

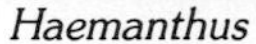
Haemanthus

Hippeastrum

Hyacinth

***Haemanthus* species**
Blood lily
This bulbous lily is fairly easy to grow and provides a splendid bloom to reward your efforts. The pink to red flowers are borne in a ball-like cluster at the end of a stem 12 to 24 inches tall. Flowering generally occurs in the early summer. The plant is large and needs room. The leaves are few in number, but can be 6 inches wide and 1 foot long.

LIGHT: Provide at least 4 hours of curtain-filtered sunlight in a bright south-, east-, or west-facing window.

WATER: Keep the plant very moist during growth and flowering. Allow it to dry between waterings at other times.

HUMIDITY: This plant requires moist air. Use a humidifier for best results.

TEMPERATURES: 50° F to 55° F nights, 65° F to 70° F days.

PROBLEMS: This plant will die back if the roots are damaged from soil that is dry or high in soluble salts.

PROPAGATION: Start new plants from the small bulblets that develop beside the parent.

GROOMING: Remove old leaves as the plant goes dormant. Give these plants plenty of room.

REPOTTING: Infrequent repotting is best for this plant, as pot-bound plants bloom more profusely. Plant with the tip of the bulb out of the soil.

FERTILIZATION: Fertilize only when growing actively or flowering.

***Hippeastrum* species and hybrids**
Amaryllis
Amaryllis bulbs are commonly sold in garden centers for forcing. The plant flowers 4 to 6 weeks after planting, so you can plan blooming times according to your planting schedule. The plants are large, and flowering stalks normally extend to 2 feet and bear clusters of flowers up to 8 inches across. Many colors are available. The flower spike appears before the leaves. After flowering, tend the foliage until September, at which time you must let the plant go dormant for a couple of months. When potting the bulb, leave half of it above the soil surface. Many indoor gardeners prefer to discard these plants after flowering and buy new bulbs each year. Second-year flowers are usually neither as large nor as beautiful.

LIGHT: Place the plant in a bright, indirectly lit south-, east-, or west-facing window.

WATER: Keep the plant evenly moist. Water thoroughly and discard drainage.

HUMIDITY: Average indoor humidity levels are satisfactory.

TEMPERATURES: 55° F to 60° F nights, 70° F to 75° F days.

PROBLEMS: Poor drainage, excessive watering, or water left in the saucer will cause root rot.

PROPAGATION: Start new plants from the small bulblets that develop beside the parent. It takes several years to produce a plant of blooming size.

GROOMING: Remove old leaves as the plant goes dormant.

REPOTTING: Repotting should be done each year in the fall.

FERTILIZATION: Do not fertilize when in flower. Fertilize lightly at other times.

Hyacinthus orientalis
Hyacinth
Hyacinths are usually purchased from florists for indoor forcing in late winter. Their fragrant blossoms come in reds, blues, or white. During flowering they can be placed almost anywhere indoors, but need bright light to flourish. To force your own, buy mature bulbs in October, pot them, and allow them to root in a cool dark spot for eight weeks or so. They should bloom two or three weeks after being brought indoors. Tend the plant until it dies back, then plant it out in the garden.

LIGHT: Provide at least 4 hours of curtain-filtered sunlight, in a bright south-, east-, or west-facing window.

WATER: Keep the plant very moist during growth and flowering. Allow it to dry between waterings at other times.

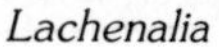

Lachenalia

Lilium longiflorum var. *eximium*

Muscari

HUMIDITY: Average indoor humidity levels are satisfactory.

TEMPERATURES: 50° F to 55° F nights, 60° F to 65° F days.

PROBLEMS: Poor drainage, excessive watering, or water left in the saucer causes root rot.

PROPAGATION: Start new plants from the small bulblets that develop beside the parent.

GROOMING: Remove old leaves as the plant goes dormant. Cut flower stalks if you wish.

REPOTTING: Plant bulb in the garden after it goes dormant.

FERTILIZATION: Do not fertilize when in flower. Fertilize lightly at other times.

Lachenalia species

Cape cowslip, leopard lily

Cape cowslips are bulbous plants that will probably be seen more frequently as indoor gardeners learn of them. They are easy to force and have striking multicolored yellow and red flowers that add color and cheer to any winter household. Most cultivars bloom in late winter if planted in the fall. The leaves are large and sometimes spotted with purple, adding interest to the plant. As with any bulb, give the plant a dormancy rest after the foliage has died down.

LIGHT: Give the plant 4 hours or more of direct sunlight in a south-facing window.

WATER: Keep the plant very moist during growth and flowering. Allow it to dry between waterings at other times.

HUMIDITY: Average indoor humidity levels are satisfactory.

TEMPERATURES: 40° F to 45° F nights, 60° F to 65° F days.

PROBLEMS: The plant will not bloom if light levels are too low. Poor drainage, excessive watering, or water left in the saucer causes root rot.

PROPAGATION: Start new plants from the small bulblets that develop beside the parent.

GROOMING: Remove old leaves as the plant goes dormant.

REPOTTING: Infrequent repotting is best for this plant.

FERTILIZATION: Fertilize only when growing actively or flowering.

Lilium species

Lily

The Easter lily (*Lilium longiflorum*) is the most common lily grown indoors and one of the most popular flowering pot plants sold in the United States. It is occasionally used as a cut flower as well. Many other species of *Lilium* can also be grown indoors. Blooming plants last longer if they are kept cool, out of drafts, and constantly moist. Remove the pollen-bearing yellow anthers just as the flower opens, to prolong its life. After flowering, care for the plant until the foliage yellows, place it in a bright window, and fertilize it lightly. Many gardeners plant lily bulbs outdoors in the garden after they have bloomed and they flower every summer thereafter. You might get another Easter flowering if you plunge the plant outdoors in its pot, protect it from early freezes, and bring it into the house in late November. Force it in a warm spot under bright light. Because Easter varies in its date, this will be difficult to do precisely.

LIGHT: This plant survives in low light, light under which you can read comfortably. Place the plant in a bright, indirectly lit south-, east-, or west-facing window after flowering is over.

WATER: Keep the plant evenly moist. Water thoroughly and discard drainage.

HUMIDITY: Average indoor humidity levels are satisfactory.

TEMPERATURES: 50° F to 55° F nights, 60° F to 65° F days.

PROBLEMS: Poor drainage, excessive watering, or water left in the saucer will cause root rot.

PROPAGATION: Bulb scales will eventually develop into bulbs after several years. Secondary bulbs can also be removed from the parent bulb.

GROOMING: Cut flower stalks if you wish.

REPOTTING: Plant outdoors after foliage dies back.

FERTILIZATION: Do not fertilize when in flower. Fertilize lightly at other times.

Muscari species

Grape hyacinth

These small bulbous plants are popular because they are easy to force into bloom. Their small size makes them ideal windowsill or desk plants. The tiny blue or white flowers are borne on 6-inch stalks. They are

Narcissus poeticus

fragrant and last for quite a while in midwinter or early spring. The leaves are narrow and grasslike, and arch outwards from the bulb tip. Force them as you do other common bulbs. Purchase mature bulbs in October, pot them and keep them cool until January, then bring in for flowering. Tend the plant until the foliage dies back and then plant in the garden.

LIGHT: Place the plant in a bright, indirectly lit south-, east-, or west-facing window.

WATER: Keep the plant evenly moist. Water thoroughly and discard drainage.

HUMIDITY: Average indoor humidity levels are satisfactory.

TEMPERATURES: 40° F to 45° F nights, 60° F to 65° F days.

PROBLEMS: No significant problems.

PROPAGATION: Start new plants from the small bulblets that develop beside the parent bulb.

GROOMING: Remove old leaves as the plant goes dormant.

REPOTTING: Transplant into the garden in the spring.

FERTILIZATION: Do not fertilize when in flower. Fertilize lightly at other times.

***Narcissus* hybrids and cultivars**
Daffodil, narcissus

Two types of narcissus bulbs are used for indoor forcing. The plants that produce a single trumpet-shaped flower on a 12-inch stalk are called daffodils. Those that produce a group of smaller flowers on a single stem are called tazettas. Many cultivars in either group are suitable for indoor gardeners. Buy mature bulbs in the fall and "plant" them in peat moss, pebbles, or sand so that one half of the bulb is out of the water or planting media. Place them at 35° F to 40° F for several weeks until they sprout about 4 inches of growth. Then bring them indoors for flowering. Keep them cool during this time and do not fertilize. Discard the plant after flowering or—if it is hardy—place it outdoors in a flower bed. Allow the leaves to die back naturally.

LIGHT: This plant will survive in low light, light by which you can read comfortably. After flowering, place in a bright window until the foliage dies back.

WATER: Keep very moist while flowering.

HUMIDITY: Average indoor humidity levels are satisfactory.

TEMPERATURES: 40° F to 45° F nights, 60° F to 65° F days.

PROBLEMS: No special problems.

PROPAGATION: Buy mature bulbs or take a large division bulb from a garden plant.

GROOMING: Discard or plant outdoors after flowering.

REPOTTING: Not necessary.

FERTILIZATION: Do not fertilize when in flower. Fertilize lightly at other times.

Ornithogalum

***Ornithogalum* species**
Star-of-Bethlehem, chincherinchee, false sea onion

These bulbous plants produce fragrant white flowers in a cluster on a long stalk. The flowers last several weeks with good care. They are easy to grow, but need ample winter light if you intend to keep them from year to year. As with any bulbous plant, give them a dormancy rest after the foliage dies back. Star-of-Bethlehem is a large plant, with leaves that are sometimes 2 feet long and flowers often 2 or 3 feet high.

LIGHT: Provide curtain-filtered sunlight in the summer in a south- or west-facing window. Give about 4 hours of direct sun in the winter.

WATER: Approach dryness before watering, but water thoroughly and discard drainage each time.

HUMIDITY: Average indoor humidity levels are satisfactory.

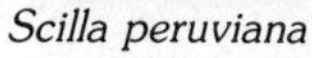

Scilla peruviana

Tulbaghia

TEMPERATURES: 50° F to 55° F nights, 65° F to 70° F days.

PROBLEMS: The plant will not bloom if light levels are too low.

PROPAGATION: Start new plants from the small bulblets that develop beside the parent.

GROOMING: Remove old leaves as the plant goes dormant.

REPOTTING: Repot in the early fall. Plant the bulb very shallowly.

FERTILIZATION: Fertilize only when growing actively or flowering.

Scilla species
Squill

Squills are commonly used outdoors as one of the earliest of the spring-flowering bulbs. The bell-like blue flowers are produced on stalks a few inches high. The moderate size of most squills makes them especially suited for winter flowering plants on a windowsill in the home. Pot several mature bulbs together in a pot in October and place in a cool (not freezing) spot until January. After the foliage declines, you can plant them in the garden.

LIGHT: Provide at least 4 hours of curtain-filtered sunlight in a bright south-, east-, or west-facing window.

WATER: Keep the plant very moist during growth and flowering. Allow it to dry between waterings at other times.

HUMIDITY: Average indoor humidity levels are satisfactory.

TEMPERATURES: 40° F to 45° F nights, 60° F to 65° F days.

PROBLEMS: No serious problems.

PROPAGATION: Start new plants from the small bulblets that develop beside the parent.

GROOMING: Cut flower stalks if you wish. Remove old leaves as the plant goes dormant. Plant in the garden after it goes dormant.

REPOTTING: Not necessary.

FERTILIZATION: Do not fertilize indoors.

Sprekelia formosissima
Aztec lily, Jacobean lily, St. James' lily

Aztec lilies are popular bulbs for indoor forcing in the spring. They will last for several years in the pot if given ample light after blooming and if allowed to rest in the fall. The leaves are not particularly attractive, but must be maintained to build the bulb for its next flowering. The plant is medium sized, with leaves up to 18 inches long. Keep the plant warm and fertilized while growing.

LIGHT: Give the plant 4 hours or more of direct sunlight in a south-facing window.

WATER: Keep moist when growing. Keep dry when dormant, but water occasionally.

HUMIDITY: Average indoor humidity levels are satisfactory.

TEMPERATURES: 60° F to 65° F nights and 70° F to 75° F days—during the growing season (February to September); 40° F to 45° F during fall and early winter.

PROBLEMS: No serious problems.

PROPAGATION: Start new plants from the small bulblets that develop beside the parent.

GROOMING: Cut flower stalks if you wish. Remove old leaves as the plant goes dormant.

REPOTTING: Repot every 3 or 4 years. Plant so that the upper third of the bulb is out of the soil.

FERTILIZATION: Fertilize only when growing actively or flowering.

Tulbaghia fragrans
Society garlic, violet tulbaghia

Although society garlic is a bulb, it will bloom repeatedly throughout the year if given lots of light, water, and fertilizer. The flowers are borne in clusters on 15-inch-tall stalks. They are usually lavender and mildly fragrant. Do not bruise or break the leaves. Broken leaves smell like garlic, which may be an odor you do not want in your indoor garden. The bulbs of society garlic multiply rapidly and require frequent division.

LIGHT: Give the plant 4 hours or more of direct sunlight in a south-facing window.

WATER: Keep the plant evenly moist. Water thoroughly and discard drainage.

HUMIDITY: Average indoor humidity levels are satisfactory.

Tulipa

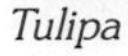

Zephyranthes candida

TEMPERATURES: 40° F to 45° F nights, 60° F to 65° F days.

PROBLEMS: The leaves smell of garlic when bruised or broken. The plant will not bloom if light levels are too low.

PROPAGATION: New plants are started by dividing an older specimen.

GROOMING: Pick off yellowed leaves as they occur.

REPOTTING: Repot each year.

FERTILIZATION: Fertilize all year, but more heavily in the summer.

Tulipa species
Tulip

Hundreds of tulip cultivars are available to gardeners. While most are suitable for indoor forcing, the smaller growing cultivars may be best suited. Tulips can be purchased already in bud from many florists. To force your own, purchase several mature bulbs in October, pot them together in one pot, and place them in a cool spot until January or February. Many devoted indoor gardeners use an old refrigerator for this purpose. Be sure and pot the bulbs with the flat side outwards to get better foliage orientation. After the foliage has died down, most tulips can be planted in the garden. The bulbs of some cultivars divide readily, but others are difficult to propagate in this manner.

LIGHT: Place anywhere during flowering. Provide at least 4 hours of curtain-filtered sunlight in a bright south-, east-, or west-facing window before and after flowering.

WATER: Keep the plant very moist during growth and flowering. Allow it to dry between waterings at other times.

HUMIDITY: Average indoor humidity levels are satisfactory.

TEMPERATURES: 50° F to 55° F nights, 60° F to 65° F days.

PROBLEMS: This plant is subject to crown rot if planted too deeply, watered over the crown, or watered late in the day.

PROPAGATION: Start new plants from the small bulblets that develop beside the parent.

GROOMING: Cut flower stalks if you wish. Remove old leaves as the plant goes dormant.

REPOTTING: Repot each year.

FERTILIZATION: Lightly fertilize after flowering.

Veltheimia species
Veltheimia, forest lily

Forest lilies are large, winter-flowering bulbous plants. The flowers, up to 10 in a cluster, are borne on a stalk that may be 2 feet high. They are light purple, tubular, and about 1 inch long. The glossy green leaves are 12 inches long and arch outward from the base of the flower stalk. Give the plants a dormancy rest in the summer.

LIGHT: Provide at least 4 hours of curtain-filtered sunlight in a bright south-, east-, or west-facing window during flowering. Give the plant 4 hours or more of direct sunlight in a south-facing window after flowering.

WATER: Keep the plant very moist during growth and flowering. Allow it to dry between waterings at other times.

HUMIDITY: Average indoor humidity levels are satisfactory.

TEMPERATURES: 50° F to 55° F nights, 60° F to 65° F days.

PROBLEMS: No serious problems.

PROPAGATION: Start new plants from the small bulblets that develop beside the parent.

GROOMING: Pick off yellowed leaves as they occur. Cut flower stalks if you wish.

REPOTTING: Infrequent repotting is best for this plant. Bulb can be repotted when growth starts in the fall.

FERTILIZATION: Fertilize only when growing actively or flowering.

Zephyranthes species
Zephyr lily

Many species and hybrids of zephyr lilies are available in the trade. Most are easy to grow and of moderate size. They bloom at various times, and often more than once in a year. Flowers can be pink, yellow, orange, or white and are borne singly on a stalk, like daffodils. The leaves are grasslike and about 1 foot long. Give the plants a rest period of 2 months after foliage has died back. Keep it in a sunny, cool spot for flowering.

Kohleria 'Flirt'

LIGHT: Give the plant 4 hours or more of direct sunlight in a south-facing window.

WATER: Keep the plant very moist during growth and flowering. Allow it to dry between waterings at other times.

HUMIDITY: Average indoor humidity levels are satisfactory.

TEMPERATURES: 40° F to 45° F nights, 60° F to 65° F days.

PROBLEMS: Poor drainage, excessive watering, or water left in the saucer causes root rot.

PROPAGATION: Start new plants from the small bulblets that develop beside the parent.

GROOMING: Cut flower stalks if you wish. Remove old leaves as the plant goes dormant.

REPOTTING: Repot each year.

FERTILIZATION: Fertilize only when growing actively or flowering.

Kohleria species

Kohleria

Kohlerias are gesneriads, and have the typically herbaceous stems and soft, hairy foliage of many members of this large family. They are easy to grow, but tend to get leggy when subjected to warm summer nights. Older cultivars grow quite tall. Well-grown, mature plants can be 2 feet tall, with stunning flowering tops, which may remind you of a bright candle. Hybridizers are working on new, more compact kohlerias that may even be attractive in hanging baskets. Winter-flowering cultivars are common, but these plants must be given lots of light. Kohlerias need a rest after flowering. Cut them back during this period. Flower colors are mostly red or yellow.

LIGHT: Provide at least 4 hours of curtain-filtered sunlight in a bright south-, east-, or west-facing window.

WATER: Keep the plant moist during growth and flowering. If you allow it to dry completely between waterings, you will force it into dormancy.

HUMIDITY: This plant requires moist air. Use a humidifier for best results.

TEMPERATURES: 50° F to 55° F nights, 60° F to 70° F days.

PROBLEMS: Plants get spindly in low light or in too much warmth.

PROPAGATION: Take cuttings from stems or shoots before they have hardened or matured. These cuttings form their own rhizomes, which will multiply in the pot, and new plants can be grown from them. Large rhizomes may even be broken in pieces, each of which will grow a new plant.

GROOMING: Prune back after flowering or fruiting.

REPOTTING: Repot after the flowering and dormancy period.

FERTILIZATION: Fertilize only when actively growing or flowering.

Sinningia hybrids

Gloxinera

Gloxineras are developed by crossing the popularly known florist's gloxinia with another *Sinningia* species. There is a lot of variation from one cultivar to another in these herbaceous plants. Most flower in late spring and summer, with clusters of red or lavender blossoms borne on stalks above the foliage. Many miniature gloxineras are suited for well-lit terrariums. Keep gloxineras moist and warm when flowering. Allow the plants to rest after flowering. If you wish, tubers can be removed, stored in a bag with some damp peat in a cool spot, and replanted when they sprout.

LIGHT: Provide at least 4 hours of curtain-filtered sunlight, in a bright south-, east- or west-facing window.

WATER: Keep the plant evenly moist. Water thoroughly and discard drainage. Terrarium conditions are ideal.

HUMIDITY: This plant requires moist air. Use a humidifier for best results.

TEMPERATURES: 65° F to 70° F nights, 75° F to 80° F days.

PROBLEMS: The plant will not bloom if light levels are too low. The leaves will scorch (turn brown and curl) if this plant is in a draft or in dry air or exposed to too much direct sunlight.

PROPAGATION: Start new plants from the small tubers that develop beside the parent.

GROOMING: Prune back after flowering.

REPOTTING: Repot in the winter or early spring as needed.

FERTILIZATION: Fertilize only when growing actively or flowering.

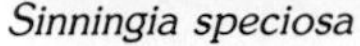

Sinningia speciosa

Anthurium scherzeranum

Sinningia speciosa
Florist's gloxinia
Florist's gloxinias are commonly available in many flower colors. The bell-shaped flowers of these gesneriads are produced in a cluster above the almost stalkless leaves. The standard cultivars can get quite large, but more compact ones are now available. Miniature gloxinias are widely grown by hobbyists. These stay very small, similar to miniature African violets. Many growers prefer to grow them in bubble bowls or very small terrariums, as these provide the best growing conditions.

LIGHT: Provide curtain-filtered sunlight in the summer in a south- or west-facing window. Give 6 or more hours of direct sun in the winter.

WATER: Water thoroughly, but allow the soil to dry out between waterings.

HUMIDITY: This plant requires moist air. Use a humidifier for best results.

TEMPERATURES: 65° F to 70° F nights, 75° F to 80° F days.

PROBLEMS: This plant is subject to crown rot if planted too deeply, watered over the crown, or watered late in the day. The leaves will scorch (turn brown and curl) if this plant is in a draft or in dry air, or in too much direct sun.

PROPAGATION: Take leaf cuttings at any time. Miniature florist's gloxinias are almost always grown from seed, as the leaves are too small to handle easily.

GROOMING: Remove old leaves as the plant goes dormant.

REPOTTING: Repot each year, after the rest period.

FERTILIZATION: Fertilize only when growing actively or flowering.

PLANTS USED INDOORS ONLY WHILE IN FLOWER

In this group of plants you will find many of the most familiar blooming potted plants. These are the plants you buy when you want flowers (and their attendant color, grace, and beauty) indoors, but desire something more than a cut flower arrangement. The idea of living, growing flowers is very appealing, and in addition, these flowers generally last longer in the home than a cut flower arrangement.

These flowering plants are grouped under this category so that you will clearly note that they are not to be thought of as permanent additions to your indoor garden. They are popular and useful because of their great beauty and spectacular appearance. They are often used as gifts, to be discarded when they have finished blooming. Some can be placed in the garden for later seasonal bloom and year-round enjoyment.

The following paragraphs and care guides will tell you something of the size, form, and color of plants often used indoors while flowering. The care instructions give information on prolonging the blooms (rather than growing the plants). Finally, information is provided for those with window boxes or greenhouses who wish to grow these plants before or after they bloom.

***Anthurium* species**
Flamingo flower, tallflower
Anthuriums are one of the most famous tropical flowering plants. The flowers are long lasting and are often used in weddings on Hawaii and other islands in the Pacific. They are popular in cut flower arrangements around the world. The red or orange portion of the "flower" is actually a bract (modified leaf) that surrounds the tiny flowers appearing on a spike or spadix. Most flamingo flowers are very large plants. *Anthurium scherzeranum* is much smaller and better suited for indoor use or greenhouse culture. Keep flamingo flowers in humid air and fertilize well when actively growing.

LIGHT: Provide at least 4 hours of curtain-filtered sunlight in a bright south-, east-, or west-facing window while in flower. Keep in a greenhouse at other times.

WATER: Keep the plant evenly moist. Water thoroughly and discard drainage.

HUMIDITY: This plant requires moist air. Use a humidifier for best results.

TEMPERATURES: 55° F to 60° F nights, 70° F to 75° F days.

PROBLEMS: The leaves will scorch (turn brown and curl) if this plant is in a draft or in dry air. The plant will not bloom if light levels are too low.

Aphelandra aurantiaca

Camellia japonica 'Dahurica Variegata'

Calceolaria herbeohybrida

PROPAGATION: Remove new plantlets or rooted side shoots as they form.

GROOMING: As high crowns form, mound up soil. Remove aerial roots.

REPOTTING: Pot with room at the top to mound up soil as crown develops.

FERTILIZATION: Fertilize three times a year, in spring, midsummer, and early fall.

Aphelandra squarrosa
Zebra plant

This attractive foliage plant produces yellow flowers in a conelike spike in the fall if it has been grown vigorously over the summer. After flowering, cut the plant back severely and keep it in a warm and well-lit place for the rest of the winter. A humidified room will be needed to prevent leaf scorch. Fertilize well in the summer to promote blooming.

LIGHT: Provide curtain-filtered sunlight in a south- or west-facing window while in bloom and for the remainder of the winter. Move outdoors or to the greenhouse during the summer until new blossoms appear.

WATER: Keep the plant evenly moist. Water thoroughly and discard drainage.

HUMIDITY: This plant requires moist air. Use a humidifier for best results.

TEMPERATURES: 65° F to 70° F nights, 75° F to 80° F days.

PROBLEMS: The leaves will scorch (turn brown and curl) if this plant is in a draft or in dry air. The leaves will drop if soil moisture is too wet or too dry.

PROPAGATION: Remove new plantlets or rooted side shoots as they form.

GROOMING: Prune back after flowering or fruiting.

REPOTTING: Cut back and repot when flowering stops.

FERTILIZATION: Fertilize only during the late spring and summer months.

Calceolaria species
Pocketbook flower, slipperwort

Pocketbook flowers are popular in Europe and are becoming increasingly available in the United States. They come in many colors from red, pink, and maroon, to yellow. Most have purple or brown markings on the petals. The plants are difficult to grow from seeds because they are sensitive to improper watering and fertilizing. They like cool nights and are well suited to a small greenhouse or window box. Pinch out stems and train them into bushy plants prior to flowering.

LIGHT: Blooming plants can be placed anywhere. Growing plants need greenhouse conditions to form flower buds.

WATER: Keep the plant evenly moist. Water thoroughly and discard drainage.

HUMIDITY: This plant requires moist air. Use a humidifier for best results.

TEMPERATURES: 40° F to 45° F nights, 60° F to 65° F days.

PROBLEMS: Whiteflies sometimes infest this plant. It is also subject to crown rot if deeply planted, watered over the crown, or watered late in the day. The leaves will scorch (turn brown and curl) if this plant is in a draft or in dry air.

PROPAGATION: Start from seeds. Begin in a small pot and transplant into larger pots, as needed.

GROOMING: Pinch out stem tips of young or regrowing plants to get better form. Stop pinching when the flowering period approaches. It is best to discard the plant after flowering.

REPOTTING: Transplant seedlings several times as the plants grow.

FERTILIZATION: Fertilize three times a year, in spring, midsummer, and early fall. Do not fertilize blooming plants.

Camellia japonica
Camellia

These popular garden shrubs of the southern and western states are available in over 2,000 known cultivars. The plants are not easy to grow indoors, but can be successfully cultured if given good light, kept cool and evenly moist, and kept out of drafts. It is

Capsicum annuum

Chrysanthemum × morifolium

especially important to guard against temperature or moisture fluctuations after bud set in the late winter. Do not fertilize in the winter, and keep in curtain-filtered sunlight. Discard the plants when they get too large.

LIGHT: Provide at least 4 hours of curtain-filtered sunlight in a bright south-, east-, or west-facing window while in flower. Move outdoors or to a greenhouse at other times.

WATER: Keep the plant evenly moist. Water thoroughly and discard drainage.

HUMIDITY: This plant requires moist air. Use a humidifier for best results.

TEMPERATURES: 40° F to 45° F nights, 60° F to 65° F days.

PROBLEMS: The leaves will scorch (turn brown and curl) if this plant is in a draft or in dry air. The leaves of this plant will drop if soil moisture is too wet or too dry.

PROPAGATION: Take cuttings from recently matured stems or shoots.

GROOMING: Prune back after flowering or fruiting.

REPOTTING: Infrequent repotting is best for this plant. Cut back and repot when flowering stops.

FERTILIZATION: Fertilize only during the late spring and summer months.

Capsicum species
Ornamental pepper

These plants are not particularly noteworthy until they become loaded with fruit in late summer and fall. Since it takes very good light to accomplish heavy blossoming and fruit set, ornamental peppers are best suited for greenhouse culture. The fruit changes from green to yellow to bright red as it matures. All stages of maturity are commonly on a plant at the same time. Even small plants will set fruit. Bring them in periodically as decorations on tables or windowsills. The fruit is edible, but it is a chili pepper and is extremely hot. Do not confuse this plant with Jerusalem cherry (*Solanum pseudocapsicum*), whose fruits are poisonous.

LIGHT: Give the plant 4 hours or more of direct sunlight in a south-facing window. This plant does best in a greenhouse setting when being brought into flower.

WATER: Water thoroughly, but allow the soil to approach dryness between waterings.

HUMIDITY: Average indoor humidity levels are satisfactory.

TEMPERATURES: 55° F to 60° F nights, 70° F to 75° F days.

PROBLEMS: The plant will not bloom if light levels are too low. The leaves of this plant will drop if soil moisture is too wet or too dry. Aphids and spider mites may infest these plants.

PROPAGATION: Start from seeds. Begin in a small pot and transplant into larger pots, as needed.

GROOMING: Prune back in the spring. Pinch out stem tips of young or regrowing plants to improve form, but do not destroy flower buds.

REPOTTING: Repot in the winter or early spring, as needed.

FERTILIZATION: Fertilize only when growing actively or flowering.

Chrysanthemum × *morifolium*
Mum, florist's mum

Mums are popular flowering houseplants sold in most garden centers and florists' shops. Although the fall is the natural blooming period for mums, the cultural practices of the modern greenhouse have made these plants available in bloom all year around. Buy mums only after the buds have begun to open, but are not yet in full bloom. To prolong their life indoors, keep them in brightly lit locations and do not overwater. In mild climates these plants can be put out in the garden after they have bloomed. In northern states most cultivars of florist's mums are not winter hardy and should be discarded after blooming. Plants purchased in the spring may bloom again in the garden in the fall. Keep the growing tips pinched out until August to create bushy, well-budded plants.

LIGHT: Place the plant in a bright, indirectly lit south-, east-, or west-facing window while in flower. To induce flowering, provide short days in a greenhouse setting.

WATER: Water thoroughly, but allow the soil to dry out between waterings.

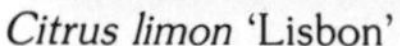

Citrus limon 'Lisbon'

Clerodendrum thomsoniae 'Variegata'

HUMIDITY: Average indoor humidity levels are satisfactory.

TEMPERATURES: 40° F to 45° F nights, 60° F to 65° F days.

PROBLEMS: Flowers will not open properly in low light. This plant is subject to infestations of spider mites and whiteflies.

PROPAGATION: Take cuttings from stems or shoots before they have hardened or matured.

GROOMING: Prune back after flowering. In northern climates it is best to discard this plant after flowering.

REPOTTING: Not needed.

FERTILIZATION: Not needed indoors.

Citrus limon
Lemon

Lemons have been grown in greenhouses in the United States for many years. The very large Italian pie lemon (fruit is 8 inches across) used to be seen routinely in greenhouses, where it provided the raw materials for the cook of the family. The 'Meyer' lemon is one of the best of the citrus for indoor gardening. It still requires lots of light, and may not bloom unless light levels are quite high. It can be pruned and kept in bounds, however. Grow in a sandy potting soil and keep well fertilized all year, especially in the summer.

LIGHT: Give the plant 4 hours or more of direct sunlight in a south-facing window. This plant does best in a greenhouse setting, but can be brought indoors temporarily.

WATER: Approach dryness before watering, but water thoroughly and discard drainage each time.

HUMIDITY: Average indoor humidity levels are satisfactory.

TEMPERATURES: 50° F to 55° F nights, 65° F to 70° F days.

PROBLEMS: The plant will not bloom if light levels are too low. The leaves will drop if soil is too wet or too dry. Mealybugs, scale, and spider mites sometimes attack this plant. Leaf yellowing occurs from soil high in soluble salts, high pH, or lack of trace elements (particularly iron).

PROPAGATION: Take stem cuttings at any time.

GROOMING: Prune to control height and thin out foliage in late winter. Keep the plant to the desired height and shape with light pruning or clipping at any time.

REPOTTING: Infrequent repotting is best for this plant.

FERTILIZATION: Use an acid-balanced fertilizer throughout the year and add trace elements once in the spring.

Clerodendrum species
Glorybower, bleeding heart

Glorybowers are actually woody shrubs that get quite large where they grow naturally outdoors. The most popular species is *Clerodendrum thomsoniae*, which has beautiful and intricate white-and-red flowers, thus spawning the name bleeding heart. It is often found in florists' shops. The flowers are borne in clusters on trailing stems, thus the plant is commonly used in hanging baskets. Keep the plants warm and give them plenty of room.

LIGHT: Provide at least 4 hours of curtain-filtered sunlight in a bright south-, east-, or west-facing window while in bloom. Grow in a greenhouse until they begin to flower.

WATER: Keep the plant very moist during growth and flowering. Allow it to dry between waterings at other times.

HUMIDITY: Average indoor humidity levels are satisfactory.

TEMPERATURES: 55° F to 60° F nights, 70° F to 75° F days.

PROBLEMS: Spider mites often attack this plant, especially if it is too dry. Poor drainage, excessive watering, or water left in the saucer causes root rot.

PROPAGATION: Take cuttings from stems or shoots before they have hardened or matured.

GROOMING: Prune back after flowering or fruiting. Pinch out stem tips of young or regrowing plants to improve form, but do not destroy flower buds.

REPOTTING: Infrequent repotting is best for this plant.

FERTILIZATION: Fertilize only during the late spring and summer months.

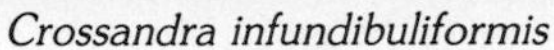

Crossandra infundibuliformis

Cyclamen persicum

Euphorbia pulcherrima

Crossandra infundibuliformis
Firecracker flower

If given ample light and fertilizer, firecracker flowers bloom almost continuously. The flowers appear on short stalks on the ends of growing shoots, overlapping one another on the budded shoot. They are a bright salmon-red color. Keep these plants very moist with frequent waterings. Do not let them stand in water. Because they are constantly growing, fertilize all year. Cut them back as needed to maintain their shape. Discard them when they get leggy or pot-bound.

LIGHT: Provide curtain-filtered sunlight in the summer in a south- or west-facing window. When flowering stops, move to a greenhouse setting. Keep in about 4 hours of direct sun in the winter.

WATER: Keep the plant evenly moist. Water thoroughly and discard drainage.

HUMIDITY: This plant requires moist air. Use a humidifier for best results.

TEMPERATURES: 55° F to 60° F nights, 70° F to 75° F days.

PROBLEMS: Poor drainage, excessive watering, or water left in the saucer causes root rot.

PROPAGATION: Start from seeds. Begin in a small pot and transplant upwards as needed. Take stem cuttings at any time.

GROOMING: Keep the plant to the desired height and shape with light pruning or clipping at any time. Start new plants and replace older specimens when they get weak.

REPOTTING: Transplant seedlings several times as the plants grow.

FERTILIZATION: Fertilize all year, but more heavily in the summer.

Cyclamen persicum
Florist's cyclamen

Florists carry blooming cyclamen plants throughout late fall and winter. Many colors are available, and some shops now sell the newer dwarf forms. Purchase a plant that has just begun to bloom. Examine the crown of the plant, checking to see if there are abundant buds coming up and that no rot is present. Keep the plant in good light, but in a cool spot. After blooming, the plant will go dormant. Put it in a cool spot after the foliage dies back and let the soil dry. In midsummer repot the tuber with new soil in a small pot and place it in a warm spot to encourage root growth. As the plant grows, gradually return it to a cool location (55° F) to induce blooming.

LIGHT: Blooming plants can be placed almost anywhere indoors. Before and after flowering, move it to the greenhouse.

WATER: Keep the plant evenly moist. Water thoroughly and discard drainage. Allow the plant to go completely dormant after flowering.

HUMIDITY: Average indoor humidity levels are satisfactory.

TEMPERATURES: 40° F to 45° F nights, 60° F to 65° F days.

PROBLEMS: This plant is subject to crown rot if planted deeply, watered over the crown, or watered late in the day.

PROPAGATION: Start new plants from the small tubers that develop beside the parent.

GROOMING: Remove old leaves as the plant goes dormant.

REPOTTING: Repot each year in late summer.

FERTILIZATION: Fertilize only when actively growing or flowering.

Euphorbia pulcherrima
Poinsettia

The poinsettia "flower" is actually made up of red bracts (modified leaves) surrounding a cluster of tiny true flowers. These Christmas plants come in white, red, pink, or variegated colors. Many new dwarf forms are available. To prolong the bloom, keep the plant warm, moderately moist, and in a well-lit location. After flowering, they make attractive foliage plants if properly

Gardenia jasminoides

Hibiscus rosa-sinensis 'Hula Girl'

pruned. Devoted indoor gardeners can get poinsettias to produce small blooms each year if they provide 12 hours of complete darkness each day from mid-September through the end of October. Even the light of the full moon is too much during these 12 hours. Turning on a small reading light for half an hour or so will also disturb the long nights needed to induce flowering.

LIGHT: Place the plant in a bright, indirectly lit south-, east-, or west-facing window while in flower. Provide short days in a greenhouse environment to initiate flowers.

WATER: Keep the plant evenly moist. Water thoroughly and discard drainage.

HUMIDITY: Average indoor humidity levels are satisfactory.

TEMPERATURES: 50° F to 55° F nights, 60° F to 65° F days. The red color will fade if the plant is too warm.

PROBLEMS: Poor drainage, excessive watering, or water left in the saucer causes root rot. This plant is subject to infestations of whiteflies.

PROPAGATION: Take cuttings from stems or shoots before they have hardened or matured.

GROOMING: Prune back in early spring. Thin out branches in summer to produce larger bracts in the coming winter.

REPOTTING: Repot each year.

FERTILIZATION: Fertilize three times a year, in spring, midsummer, and early fall.

Gardenia jasminoides
Gardenia, Cape jasmine

These evergreen shrubs produce an abundance of extremely fragrant flowers, usually in the summer and fall. Good light, cool nights, and plenty of water are required for proper flowering. The foliage is sufficiently attractive to make the plants valued additions to an indoor garden, even if they do fail to bloom after their initial purchase. The flowers last well as cut blooms in a vase or floating in water.

LIGHT: This plant does best in a greenhouse setting. It can be brought indoors while in bloom. Keep it in a brightly lit spot.

WATER: Keep the plant very moist at all times, but do not allow it to stand in water.

HUMIDITY: This plant requires moist air. Use a humidifier for best results.

TEMPERATURES: 50° F to 55° F nights, 60° F to 65° F days.

PROBLEMS: The leaves of this plant will drop if soil is too wet or too dry. Flower buds will drop if the plant is stressed in any way.

PROPAGATION: Take cuttings from stems or shoots before they have hardened or matured.

GROOMING: Pinch out stem tips of young or regrowing plants to improve form, but do not destroy flower buds.

REPOTTING: Infrequent repotting is best for this plant.

FERTILIZATION: Fertilize all year, but more heavily in the summer. Use an acid-balanced fertilizer and add trace elements once in the spring.

Hibiscus species
Hibiscus

Hibiscuses are woody shrubs popular in outdoor landscapes of southern states. If given plenty of light and kept pruned to about 3 feet, they are easy to grow and attractive indoor plants. The plants have large blooms that are available in pink, red, yellow, orange, or white. Several flower forms are available (singles, doubles, and so forth). Although individual hibiscus flowers are short-lived, the plant blooms throughout the year.

LIGHT: Give the plant 4 hours or more of curtain-filtered sunlight in a south-facing window while in flower. Move it to the greenhouse when flowering ceases.

WATER: Keep the plant evenly moist. Water thoroughly and discard drainage.

HUMIDITY: This plant requires moist air. Use a humidifier for best results.

TEMPERATURES: 55° F to 60° F nights, 70° F to 75° F days.

PROBLEMS: Spider mites often attack this plant, especially if it is too dry. The leaves will scorch (turn brown and curl) if this plant is in a draft or in dry air and the plant will not bloom if light levels are too low.

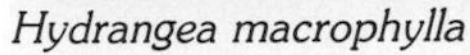

Hydrangea macrophylla

Kalanchoe

PROPAGATION: Take cuttings from stems or shoots before they have hardened or matured.

GROOMING: Keep the plant to the desired height and shape with light pruning or clipping at any time. Give these plants plenty of room.

REPOTTING: Repot in the winter or early spring, as needed.

FERTILIZATION: Fertilize all year, but more heavily in the summer.

Hydrangea macrophylla
Big-leaf hydrangea

These blooming plants are popular at Eastertime. The many flowers are borne in large ball-like clusters up to 12 inches across. The plant is a shrub that grows quite large, with shiny leaves sometimes 6 inches long. It is easy to care for while flowering, but will usually not bloom indoors a second time. To prolong the bloom, provide cool, well-lit conditions. The cultivars are usually not hardy outdoors, so big-leaf hydrangeas are usually discarded after flowering.

LIGHT: Place the plant in a bright, indirectly lit south-, east-, or west-facing window while in bloom.

WATER: Keep the plant very moist at all times, but do not allow to stand in water.

HUMIDITY: Average indoor humidity levels are satisfactory.

TEMPERATURES: 50° F to 55° F nights, 65° F to 70° F days.

PROBLEMS: The leaves will scorch (turn brown and curl) if this plant is in a draft or in dry air.

PROPAGATION: Take cuttings from stems or shoots before they have hardened or matured.

GROOMING: It might be simplest to discard this plant after flowering.

REPOTTING: Infrequent repotting is best for this plant.

FERTILIZATION: Do not fertilize when in flower. Fertilize lightly at other times. Use an acid-balanced fertilizer and add trace elements once in the spring.

Kalanchoe blossfeldiana
Kalanchoe

These pot plants bloom in winter and are becoming widely available in florists' shops and garden centers across the country. Many cultivars are available, including dwarf types, in a variety of bright colors. The tiny flowers are borne in abundance on short stalks above the thick, fleshy leaves. To prolong the bloom, purchase plants when they are just beginning to flower and keep them in brightly lit locations. Many gardeners discard the plants after flowering ends. It is difficult to promote a second season of winter flowering unless you provide 14 hours of complete darkness every night in September and October.

LIGHT: Blooming plants can be placed almost anywhere indoors. Place in a greenhouse prior to blooming.

WATER: Keep the plant evenly moist. Water thoroughly and discard drainage.

HUMIDITY: Average indoor humidity levels are satisfactory.

TEMPERATURES: 50° F to 55° F nights, 65° F to 70° F days.

PROBLEMS: Poor drainage, excessive watering, or water left in the saucer causes root rot. Some cultivars contract powdery mildew if kept in cold drafts.

PROPAGATION: Start from seeds. Begin in a small pot and transplant into larger pots as needed. Take stem cuttings at any time.

GROOMING: It is best to discard this plant after flowering.

REPOTTING: Not usually needed.

FERTILIZATION: Fertilize only when growing actively or flowering.

Primula malacoides

Primula × polyantha

Rosa

Rhododendron

Primula **species**
Primrose

Primroses produce magnificent clusters of flowers that are held on stalks above a rosette of light green leaves. Cultivars are available in red, yellow, pink, orange, blue, white, or as bicolors. It takes a lot of light, and moist air with very cool nights, to get these plants to flower properly in the winter. Thus they are usually purchased already in bloom. They can be started and grown in home greenhouses, provided they are kept moist. Any stress will lead to spider mite infestation.

Four species especially suited to indoors are *Primula malacoides* (fairy primrose), *Primula obconica* (German primrose), *Primula × polyantha* (polyantha primrose), and *Primula sinensis* (Chinese primrose).

LIGHT: Place plants anywhere while in flower, except for *Primula obconica*, which cannot take direct sun while flowering. Plants do best in a greenhouse setting before and after flowering.

WATER: Keep the plant evenly moist. Water thoroughly and discard drainage.

HUMIDITY: This plant requires moist air. Use a humidifier for best results.

TEMPERATURES: 40° F to 45° F nights, 60° F to 65° F days.

PROBLEMS: Spider mites often attack this plant, especially if it is too dry. The leaves will scorch (turn brown and curl) if this plant is in a draft or in dry air. This plant will die back if the roots are damaged from soil that is dry or high in soluble salts.

PROPAGATION: Start from seeds. Begin in a small pot and transplant into larger pots, as needed.

GROOMING: Pick off yellowed leaves as they occur.

REPOTTING: Repot in the winter or early spring, as needed.

FERTILIZATION: Do not fertilize when in flower. Fertilize lightly at other times.

Rhododendron **species**
Azalea, rhododendron

Many cultivars of azaleas and a few rhododendrons are available from florists for use as blooming houseplants. The blooms are long lasting. After flowering, many cultivars are well suited as indoor foliage plants. Some can be planted in the garden after flowering. Most of the azaleas sold in this way are not hardy in northern gardens, however. It is difficult to get them to flower in subsequent years indoors. They need greenhouse conditions to set flower buds in the fall and must have several weeks of cool, dormant rest.

LIGHT: Plants may be placed anywhere when in bloom. Afterwards, provide a greenhouse setting.

WATER: Keep the plant evenly moist. Water thoroughly and discard drainage.

HUMIDITY: This plant requires moist air. Use a humidifier for best results.

TEMPERATURES: 55° F to 60° F nights, 70° F to 75° F days.

PROBLEMS: The leaves will scorch (turn brown and curl) if this plant is in a draft or in dry air. Spider mites often attack this plant, especially if it is too dry.

PROPAGATION: Take cuttings from stems or shoots before they have hardened or matured.

GROOMING: Prune back after flowering or fruiting.

REPOTTING: Infrequent repotting is best for this plant.

FERTILIZATION: Fertilize only when growing actively or flowering. Use an acid-balanced fertilizer and add trace elements once in the spring.

Senecio × hybridus

Solanum pseudocapsicum

PROPAGATION: Start from seeds. Begin in a small pot and transplant into larger pots, as needed.

GROOMING: It is best to discard this plant after flowering.

REPOTTING: Transplant seedlings several times as the plants grow. Infrequent repotting of mature plants is best.

FERTILIZATION: Do not fertilize when in flower; fertilize lightly at other times.

***Rosa* species**
Rose (miniature)

Miniature roses need lots of light to do well indoors. Many cultivars in a variety of flower colors are available. Many produce fragrant blooms. Prune them frequently to maintain the desired shape and size. Excessive dryness may lead to spider mite infestations. Cold drafts often bring on powdery mildew disease.

LIGHT: Provide curtain-filtered sunlight in the summer in a south- or west-facing window. Give about 4 hours of direct sun in the winter. As plants become spindly and cease to flower well, place them in a greenhouse setting.

WATER: Keep the plant evenly moist. Water thoroughly and discard drainage.

HUMIDITY: Average indoor humidity levels are satisfactory.

TEMPERATURES: 50° F to 55° F nights, 60° F to 65° F days.

PROBLEMS: Spider mites often attack this plant, especially if it is too dry. The plant will not bloom if light levels are too low. The leaves will drop if soil is too wet or too dry.

PROPAGATION: Take stem cuttings at any time.

GROOMING: Keep the plant to the desired height and shape with light pruning or clipping at any time.

REPOTTING: This plant can be repotted at any time.

FERTILIZATION: Fertilize all year, but more heavily in the summer.

Senecio × hybridus
Cineraria

Cinerarias are popular winter-blooming plants that many florists now stock regularly. They are easy to grow from seed, and produce a large cluster of flowers in a variety of colors from pink to dark blue. Some of the hybrid seedlings have dark foliage with a purplish cast when viewed from below. It is interesting to start with a seed mixture and see what different forms and flower colors you get. Keep the plants cool when flowering. Give them plenty of light prior to flowering so they will not get too leggy.

LIGHT: Provide at least 4 hours of curtain-filtered sunlight in a bright south-, east-, or west-facing window while in bloom. Grow in a greenhouse before flowering.

WATER: Keep the plant evenly moist. Water thoroughly and discard drainage.

HUMIDITY: Average indoor humidity levels are satisfactory.

TEMPERATURES: 45° F to 50° F nights, 60° F to 65° F days, or cooler during flowering.

PROBLEMS: This plant is subject to infestations of whiteflies, aphids, and spider mites. Powdery mildew can also be a problem.

Solanum pseudocapsicum
Jerusalem cherry, Christmas cherry

Jerusalem cherries are very popular houseplants for the Christmas season because they bear red fruit at this time of the year if given enough light. They also bear blossoms and new green fruits that turn orange and finally red as they mature. The fruits are poisonous.

LIGHT: Provide at least 4 hours of sunlight from a bright south-, east-, or west-facing window.

WATER: Keep evenly moist. Water thoroughly and discard drainage.

HUMIDITY: This plant requires moist air. Use a humidifier for best results.

TEMPERATURES: 50° F to 55° F nights, 60° F to 65° F days.

PROBLEMS: The foliage has an odor that some find objectionable. Spider mites often attack this plant.

PROPAGATION: Start from seeds. Begin in a small pot and transplant into larger pots, as needed.

GROOMING: Pinch tips of young or regrowing plants to improve form, but do not destroy flower buds. Discard plants when they get too leggy.

REPOTTING: Transplant seedlings several times as the plants grow. Repot in early winter or spring, as needed.

FERTILIZATION: Fertilize only when plant is actively growing or flowering.

Index

NOTE: Encyclopedia pages appear in **boldface**; photographs in *italics*.

U.S. Measure and Metric Measure Conversion Chart

		Formulas for Exact Measures			Rounded Measures for Quick Reference		
	Symbol	**When you know:**	**Multiply by**	**To find:**			
Mass (Weight)	oz	ounces	28.35	grams	1 oz		= 30 g
	lb	pounds	0.45	kilograms	4 oz		= 115 g
	g	grams	0.035	ounces	8 oz		= 225 g
	kg	kilograms	2.2	pounds	16 oz	= 1 lb	= 450 kg
					32 oz	= 2 lb	= 900 kg
					36 oz	= 2 1/4 lb	= 1000g (a kg)
Volume	tsp	teaspoons	5.0	milliliters	1/4 tsp	= 1/24 oz	= 1 ml
	tbsp	tablespoons	15.0	milliliters	1/2 tsp	= 1/12 oz	= 2 ml
	fl oz	fluid ounces	29.57	milliliters	1 tsp	= 1/6 oz	= 5 ml
	c	cups	0.24	liters	1 tbsp	= 1/2 oz	= 15 ml
	pt	pints	0.47	liters	1 c	= 8 oz	= 250 ml
	qt	quarts	0.95	liters	2 c (1 pt)	= 16 oz	= 500 ml
	gal	gallons	3.785	liters	4 c (1 qt)	= 32 oz	= 1 l
	ml	milliters	0.034	fluid ounces	4 qt (1 gal)	= 128 oz	= 3 3/4 l
Length	in.	inches	2.54	centimeters	3/8 in.	= 1 cm	
	ft	feet	30.48	centimeters	1 in.	= 2.5 cm	
	yd	yards	0.9144	meters	2 in.	= 5 cm	
	mi	miles	1.609	kilometers	2-1/2 in.	= 6.5 cm	
	km	kilometers	0.621	miles	12 in. (1 ft)	= 30 cm	
	m	meters	1.094	yards	1 yd	= 90 cm	
	cm	centimeters	0.39	inches	100 ft	= 30 m	
					1 mi	= 1.6 km	
Temperature	°F	Fahrenheit	5/9 (after subtracting 32)	Celsius	32°F	= 0°C	
					68°F	= 20°C	
	°C	Celsius	9/5 (then add 32)	Fahrenheit	212°F	= 100°C	
Area	in.^{2}	square inches	6.452	square centimeters	1 in.^{2}	= 6.5 cm^2	
	ft^2	square feet	929.0	square centimeters	1 ft^2	= 930 cm^2	
	yd^2	square yards	8361.0	square centimeters	1 yd^2	= 8360 cm^2	
	a	acres	0.4047	hectares	1 a	= 4050 m^2	